MATH FOR WASTEWATER TREATMENT OPERATORS GRADES 1 AND 2

MATH FOR WASTEWATER TREATMENT OPERATORS GRADES 1 AND 2

A Guide to Preparing for Wastewater Treatment
Operator Certification Exams

John Giorgi

FIRST EDITION

This work is dedicated to my wife, Flora Zhou Giorgi; my children, Sara, Stephanie, and Steve; my mother, Thelma Giorgi; and my father, Albert Peter Giorgi.

Math for Wastewater Treatment Operators Grades 1 and 2
Copyright © 2009 American Water Works Association

AWWA Publications Manager: Gay Porter De Nileon
Technical Editor/Project Manager: Martha Ripley Gray
Production: Hop-To-It Design Works & SquareOne Publishing

Disclaimer

The authors, contributors, editors, and publisher do not assume responsibility for the validity of the content or any consequences of their use. In no event will AWWA be liable for direct, indirect, special, incidental, or consequential damages arising out of the use of information presented in this book. In particular, AWWA will not be responsible for any costs, including, but not limited to, those incurred as a result of lost revenue. In no event shall AWWA's liability exceed the amount paid for the purchase of this book.

Although this study guide has been extensively reviewed for accuracy, there may be an occasion to dispute an answer, either factually or in the interpretation of the question. Both AWWA and the author have made every effort to correct or eliminate any questions that may be confusing or ambiguous. If you do find a question that you feel is confusing or incorrect, please contact the AWWA Publishing Group.

Additionally, it is important to understand the purpose of this study guide. It does not guarantee certification. It is intended to provide the operator with an understanding of the types of math questions he or she will be presented with on a certification exam and the areas of knowledge that will be covered. AWWA highly recommends that you make use of the additional resources listed at the end of this study guide and any other resources recommended by your state certification board in preparing for your exam.

Library of Congress Cataloging-in-Publication Data

Giorgi, John.
  Math for wastewater treatment operators grades 1 and 2 : practice problems to prepare for wastewater treatment operator certification exams / John Giorgi and prepared by the editors of American Water Works Association. -- 1st ed.
    p. cm.
  Includes bibliographical references.
  ISBN 978-1-58321-587-6
  1. Sewage--Purification--Mathematics. 2. Sewage--Purification--Problems, exercises, etc. 3. Sewage disposal plants--Employees--Certification. 4. Water--Purification--Mathematics. 5. Water--Purification--Problems, exercises, etc. 6. Water treatment plants--Employees--Certification. 7. Engineering mathematics--Formulae. I. Title. II. Title: Math for wastewater treatment operators grades one and two.

  TD745.G56 2008
  628.301'51--dc22

                2008016635

American Water Works Association

6666 West Quincy Avenue
Denver, Colorado 80235-3098
800.926.7337

CONTENTS

TABLES

FIGURES

ACKNOWLEDGMENTS

I would like to thank the staff and editors of the American Water Works Association, publications manager Gay Porter De Nileon, editor Martha Ripley Gray, and reviewers Radenko Odzakovic and Tim McCandless for their help and guidance in making this book possible. Their assistance is greatly appreciated.

I am grateful to my wife, Flora, and my children, Steve, Stephanie, and Sara. Their patience and support for my long hours working on this book will always be greatly appreciated.

PREFACE

The first edition of *Math for Wastewater Treatment Operators Grades 1 and 2* was written to provide students and operators with examples of a variety of different problems that will be encountered both on certification exams and on the job. This book is divided into four main parts: the introduction, which is a review of significant numbers and rounding; two chapters on math followed by practice tests, one after each math chapter; and appendices. The math problems in chapter two are a little more difficult than the math problems in chapter one and also contain a few new types.

Each problem is presented with easily followed steps and comments to facilitate understanding. One possible way to go through the math problems presented in this study guide is for you to cover the page you are working on with a piece of paper or cardboard. Then, slowly move the cover down until you can read the question. Do the problem on a separate piece of paper. Uncover the worked solution and compare your method and result to the book's method and result. If your answer is the same, but your method is different, that's okay. Remember that there may be more than one way to solve a problem. If there is a certain problem that gives you trouble, try to do the problem again on another day until you completely understand it. Do similar problems that may be found in the other grade in the book. The more math problems you do, the more comfortable you will become with them.

Included after each grade level are tests to help you determine where your strengths or weaknesses are. Each test consists of randomly chosen problems from the associated problems in that chapter. The questions in the test are followed immediately by each individual question, procedure, and result. You can complete all the problems in the test on a separate piece of paper. After completing the test, the procedures and answers can be checked against the provided procedures and solutions.

Common conversion factors are included in appendix A for reference purposes and for doing problems in this book. Appendix B is a summary of the wastewater treatment equations. Appendix C consists of chemistry tables that are needed for some of the problems. Appendix D is the depth-to-diameter table for calculating flow in a pipeline that is not full. Appendix E contains flow charts of wastewater treatment processes. And appendix F lists abbreviations used in this book.

Any suggestions for improving this math book, including additional types of problems, would be appreciated by the author. Please send your suggestions or questions to John Giorgi at aujourney@hotmail.com, or in care of Publications Manager, AWWA, 6666 W. Quincy Avenue, Denver, CO 80235-3098.

INTRODUCTION

SIGNIFICANT FIGURES

When you see an answer to a mathematical problem, laboratory test result, or other measured values, do you ever wonder how accurate they are? The accuracy of any answer is based on how accurate the values are in determining the answer, or the accuracy of the laboratory result depends on the precision of the measuring instruments, and even the laboratory analyst.

The following discussion will show how to determine the number of significant figures or digits an answer to any particular problem should have, i.e., how many decimal places, if any, should the answer have.

The number 30.03 has four significant figures, while the number 33,000 has only two but could have three, four, or five depending on whether or not any of the zeroes are measured values. Why is this so? The number 30.03 has been measured to the hundredth place, so the zeroes that are straddled by the threes are significant. In fact, all figures to the left of a decimal point are significant (for example, 2.000 has four significant figures). The second number, 33,000, has only two because the zeroes in this case are only placeholders and are thus not significant. See the exercise below on significant figures.

"Rounding Off"

"Rounding off" numbers is simply the dropping of figures starting on the right until the appropriate numbers of significant figures remain. Let's look at the three rules and an example for each that governs the process of rounding numbers.

1. When a figure less than five is dropped, the next figure to the left remains unchanged. Thus, the number 11.24 becomes 11.2 when it is required that the four be dropped.

2. When the figure is greater than five that number is dropped and the number to the left is increased by one. Thus 11.26 will become 11.3.

3. When the figure that needs to be dropped is a five, round to the nearest even number. This prevents rounding bias. Thus 11.35 becomes 11.4 and 46.25 becomes 46.2.

The Significance or Insignificance of Zero

A zero may be a significant figure, if it is a measured value, or be insignificant and serve only as a spacer for locating the decimal point. If a zero or zeroes are used to give position value to the significant figures in the number, then the zero or zeroes are not significant. An example of this would be the following expression: 1.23 mm = 0.123 cm = 0.000123 m = 0.00000123 km. The zeroes are insignificant and only give the significant figures, 123, a position that dictates their value.

Addition and Subtraction

In addition and subtraction, only similar units and written to the same number of decimal places may be added or subtracted. Also, the number with the least decimal places, and not necessarily the fewest number of significant figures, places a limit on the number that the sum can justifiably carry, for example, when adding the following numbers: 446 mm + 185.22 cm + 18.9 m.

First, convert the quantities to similar units, which in this case will be the meter (second row below). Next, choose the least accurate number, which is 18.9. It has only one number to the right of the decimal, so the other two values will have to be rounded off (third row below).

$$
\begin{aligned}
446\,\text{mm} &= 0.446 &&= 0.4\,\text{m} \\
185.22\,\text{cm} &= 1.8522\,\text{m} &&= 1.8\,\text{m} \\
18.9\,\text{m} &= 18.9\,\text{m} &&= \underline{18.9\,\text{m}} \\
& && \ 21.1\,\text{m}
\end{aligned}
$$

When adding numbers (including negative numbers), the rule is that the least accurate number will determine the number reported as the sum. In other words, the number of significant figures reported in the sum cannot be greater than the least significant figure in the group being added. Another example is given below in which the least accurate number, 170, dictates how the other three numbers will have to be changed before addition is done.

$$
\begin{aligned}
1.023\,\text{grams (g)} &= & 1\,\text{g} \\
23.22\,\text{g} &= & 23\,\text{g} \\
170\,\text{g} &= & 170\,\text{g} \\
\underline{1.008\,\text{g}} &= & \underline{1\,\text{g}} \\
195.251\,\text{g} & & 195\,\text{g}
\end{aligned}
$$

However, you cannot report either of these values. The third value, 170 g, has two significant numbers, while all the others have four. The limiting factor is this third value, 170 g. The number 195 has three significant figures, and thus, cannot be used either. The answer must be reported as 200 g even though this looks wrong because it appears to have only one significant figure!

Multiplication and Division

In multiplication or division, the number that has the fewest significant figures will dictate how the answer will be written. Suppose we had a problem where we had to multiply two numbers: (23.88)(7.2) = 171.936. The first number has four significant figures, while the second has only two. The answer should only be written with two significant figures, as 170, because one of the numbers, 7.2, has only two significant figures. In both multiplication and division, "rounding off" never should be done before the mathematical exercise. Only the result should be appropriately "rounded off."

However, the above explanation of rounding multiplication and division problems is really an approximation of the exact rule. The exact rule states that the fractional or percentage error of a multiplication or division problem cannot be any less than the fractional or percentage error of any one factor. The exact rule always has to consider the amount of error that would result when rounding and applies to problems that have only one or two significant figures. These problems may require an additional significant number. An example follows: (0.93)(1.23) = 1.14. If this were rounded to 1.1 (two significant figures as is 0.93), as per the generalized or approximate rule, it would result in an error of approximately 3.5%, i.e., from the 1.14 answer. If the least significant number in the problem, 0.93, were written as 0.93 ± 0.01, it would result in an error of just over 1%, i.e., if it were really 0.94 or 0.92.

$$\frac{(0.01)(100\%)}{0.93} = 1.075\%, \text{ round to } 1.1\%$$

Thus, the best answer would be 1.14 because it introduces less error than the least significant number, if again it were measured wrong and were a little more or less (0.94 or 0.92).

Because this is much more difficult than what we need in this book, we will use the approximate rounding rule throughout this book, except where indicated. I stated it here so that you would be aware of this rule in case you are not already.

In the following exercise, give the number of significant figures from each of the values below:

VALUE	ANSWER
a. 8.34 lb/gal	3
b. 0.03 ntu	1
c. 19.08 mgd	4
d. 3 1-ton sulfur dioxide containers	1 or infinite*
e. 2.30 mg/L	3
f. 0.00000254	3
g. 80,000 pennies	1 to 5**
h. 0.006700	4
i. 43,560 ft³/acre-ft	4 or 5 or infinite#
j. 220 m	2 or 3
k. 9.02 mg	3
l. 10,200,050 gal	7 or 8
m. 1,000,000/mil	7 or infinite#
n. 1,440 min/day	4 or infinite+
o. 7.481 gal/ft³	4
p. 86,400 sec/day	5 or infinite+

* Because one would not divide them up in little pieces.

** Because one could argue that 1 or all of the zeroes could be significant, but not likely.

By definition, so one could extend the zeros past the decimal point to infinity.

+ By definition, these are the minutes and seconds, respectively, that are in a day; until the earth's rotational speed slows down enough to warrant a change in these times, I will call these as indicated.

See chapter 1, Significant Figures, for more practice.

ROUNDING IN THIS BOOK

The problems in this book are done in steps so the student can see each operation visually, which facilitates understanding. In so doing, a certain amount of "pre-rounding" has to occur, otherwise the numbers continue in most cases to absurdity. This "pre-rounding" was done as much as possible such that the final answer would not be affected. In general, keep at least one more significant digit in intermediate results, when doing multistep calculations, than is required in the final answer.

SET UP OF PROBLEMS IN THIS BOOK

Note that many of the same problem types have the equation written each time and that the units are written throughout each problem. The reason this is done is to help the student form a good habit, because carrying units throughout a calculation will assist in identifying which units require conversion and as a check on the final result. This habit should be carried over during a certification test or on the job. In some cases, partial credit will be given for a correct written equation. When the units are written down, it will not only help the student in setting up the problem correctly by seeing how the units will cancel, but will also make it easier for the person correcting the test to see your intent and also in correcting the problems.

Pi (π)

The number used for pi (π) will be 3.14 throughout this book.

ppm and mg/L

Since in most cases mg/L = ppm, please note that mg/L will cancel out units that are in millions. For example, mg/L will cancel with the mil (million) in mil gal.

DERIVATION OF THE NUMBER 0.785

The number 0.785 is used extensively in this book in conjunction with the diameter squared. Examples include the determination of the area of a circular reservoir or the volume of the tank. The same answer can be achieved using πr^2, where "r" equals the radius. But how is this number, 0.785, derived? The following applies:

1. **(0.785)(Diameter)^2 = πr^2**
2. **(Diameter)^2 = 4r^2**

Proof: Assume x is the number 0.785 but is not yet known. We know some number, x, times the diameter squared equals π times the radius squared. The equation is:

x(Diameter)^2 = πr^2
From 1 above and substituting x for 0.785, substitute $4r^2$ for (Diameter)^2 from 2 above.
$x(4r^2) = \pi r^2$ Rearrange equation to solve for x.

$$x = \frac{\pi r^2}{4r^2} = \frac{\pi}{4} = \frac{3.14}{4} = \mathbf{0.785}$$

Thus, x is equal to 0.785, which is what we wanted to prove. Most water treatment operators like to use (0.785)(Diameter)^2, while engineers and scientists like to use πr^2. Because both will be encountered, it is advisable to know both methods.

CHAPTER 1

WASTEWATER TREATMENT
Grade 1

Students preparing for Grade 2 through Grade 4 wastewater treatment certification tests should also understand these problems.

SIGNIFICANT FIGURES

These problems are a good refresher for the student or operator. The importance of significant figures is to not place too much importance in numbers that are insignificant and therefore meaningless.

Note: L = liter, gal = gallons, mgd = million gallons per day, g = grams, cm = centimeters

1. Addition:

a.		b.		c.	
6.071	L	488.0	gal	6.004	mgd
2.463	L	1.943	gal	0.2896	mgd
5.8	L	9.4	gal	14.2	mgd
0.158	L	23.007	gal	0.847	mgd
14.492	L	522.35	gal	21.3406	mgd

Answers:
a. **14.5 L**   b. **522.4 gal**   c. **21.3 mgd**

2. Subtraction:

a.
$$\begin{array}{r} 0.068902 \text{ g} \\ -0.0376 \quad \text{ g} \\ \hline 0.031302 \text{ g} \end{array}$$

b.
$$\begin{array}{r} 422 \quad \text{cm} \\ -5.23 \text{ cm} \\ \hline 416.77 \text{ cm} \end{array}$$

c.
$$\begin{array}{r} 1,437.5 \text{ gal} \\ -12.96 \text{ gal} \\ \hline 1,424.54 \text{ gal} \end{array}$$

Answers:
a. **0.0313 g**
b. **417 cm**
c. **1,424.5 gal**

3. Multiplication:

a. $(24.03)(0.2876) = 6.911028$, round to **6.911**

b. $(8.1)(1.6238) = 13.15278$, round to **13**

c. $(3.214 \times 10^4)(4.2 \times 10^7) = 13.4988 \times 10^{11}$, round to **13 × 10¹¹**
   (Always add the exponents $10^4 + 10^7 = 10^{11}$ or in the case when one of the exponents is negative, $10^5 + 10^{-7} = 10^{-2}$)

4. Division:

a. $\dfrac{21.973}{58.25} = \mathbf{0.3772}$ b. $\dfrac{1,402}{75.2} = \mathbf{18.6}$ c. $\dfrac{0.8334}{2.5} = \mathbf{0.33}$

COMMON MISCELLANEOUS CONVERSION PROBLEMS

These problems are a good refresher for the student or operator because they are used constantly in wastewater calculations.

5. How many pounds (lb) does exactly 2 gallons of water weigh?

Know: 1 gal = 8.34 lb

Thus: Number of lb = (8.34 lb/gal)(2 gal) = **16.68 lb**

6. How many pounds are there in 1 cubic foot (ft³) of water?

Know: 1 ft³ = 7.48 gal

Number of lb/ft³ = (8.34 lb/gal)(7.48 gal/ft³) = 62.3832 lb/ft³, round to **62.4 lb/ft³**

7. How much does exactly 25 gallons of water weigh?

Equation: **Number of lb = (Number of gal)(8.34 lb/gal)**

Substitute values and solve.

Number of lb = (25 gal)(8.34 lb/gal) = 208.5 lb, round to **208 lb**

Note: In this case 25 gal is not a measurement, but an exact amount asked for in the problem. It therefore has an infinite number of significant figures. The limiting factor in this problem is 8.34 lb/gal (three significant figures).

8. How many gallons are there in 34.8 ft³?

Know: 1 ft³ = 7.48 gal

Thus: Number of gal = (7.48 gal/ft³)(34.8 ft³) = 260.304 gal, round to **260 gal**

9. Convert 249,473 ft³ to acre-ft.

Know: 1 acre-ft = 43,560 ft³

Thus: Number of acre-ft = $\dfrac{249,472 \text{ ft}^3}{43,560 \text{ ft}^3/\text{acre--ft}}$ = **5.7271 acre-ft**

10. How many acre feet (acre-ft) are there in 4.35 million gallons (mil gal)?

Know: 1 acre of ft^3 = 43,560 ft^3 and 1 ft^3 = 7.48 gal

First, convert 4.35 million gallons to gallons.

Number of gallons = (1,000,000/mil)(4.35 mil gal) = 4,350,000 gal

Number of acre-ft = $\dfrac{(1\,\text{acre ft})\,(1\,\text{ft}^3)\,(4,350,000\,\text{gal})}{(43,560\,\text{ft}^3)\,(7.48\,\text{gal})}$ = 13.35 acre-ft, round to **13.4 acre-ft**

11. Convert 250 gallons per minute (gpm) to cubic feet per second (ft^3/s).

Know: 1 ft^3 = 7.48 gal

Thus: **Number of ft^3/s** = $\dfrac{\text{Number of gpm}}{(60\,\text{s/min})\,(7.48\,\text{gal/ft}^3)}$

Number of ft^3/s = $\dfrac{250\,\text{gpm}}{(60\,\text{s/min})\,(7.48\,\text{gal/ft}^3)}$ = 0.557 ft^3/s, round to **0.56 ft^3/s**

12. Convert 485.6 grams (g) to pounds (lb).

Know: 1 pound = 454 grams

Number of lb = $\dfrac{(1\,\text{lb})\,(485.6\,\text{g})}{454\,\text{g}}$ = 1.069 lb, round to **1.07 lb**

13. Convert 77 gallons to liters (L).

Know: 1 gallon = 3.785 liters

Thus: Number of liters = (77 gal)(3.785 L/1 gal) = 291.445 L, round to **290 L**

14.   Convert 18.2 ft³/s to million gallons per day (mgd).

Know: 1 ft³/s = 0.6463 mgd

Thus: Number of mgd = $\dfrac{(18.2 \text{ ft}^3/\text{s})(0.6463 \text{ mgd})}{1 \text{ ft}^3/\text{s}}$ = 11.7627 mgd, round to **11.8 mgd**

15.   A tank holds 500 gallons. How many cubic feet is this?

Know: 1 ft³ = 7.48 gal

Thus: Number of ft³ = $\dfrac{(500 \text{ gal})(1 \text{ ft}^3)}{7.48 \text{ gal}}$ = 66.845 ft³, round to **70 ft³**

16.   How many million gallons are there in 326 acre-ft?

Know: 43,560 ft³ = 1 acre-ft and 7.48 gal = 1 ft³

Thus: $\dfrac{(326 \text{ acre-ft})(43{,}560 \text{ ft}^3/\text{acre-ft})(7.48 \text{ gal/ft}^3)(1 \text{ mil gal})}{1{,}000{,}000}$ = 106.22 mil gal, round to **106 mil gal**

17.   Convert 16.3 ft³/s to gallons per minute (gpm).

Know: 7.48 gal = 1 ft³

Thus: (16.3 ft³/s)(7.48 gal/ft³)(60 s/min) = 7,315.44 gpm, round to **7,320 gpm**

18.   Convert 9.2 ft³/s to mgd.

Know: 7.48 gal = 1 ft³

Thus: Number of mgd = (9.2 ft³/s)(7.48 gal/ft³)(86,400 s/d)(1 mil gal/1,000,000 gal)

Number of mgd = 5.946 mgd, round to **5.9 mgd**

19. Convert 72.35 acre-ft to cubic feet.

Know: $43,560 \text{ ft}^3 = 1 \text{ acre-ft}$

Thus: $(72.35 \text{ acre-ft})(43,560 \text{ ft}^3/\text{acre-ft}) = 3,151,566 \text{ ft}^3$, round to **3,152,000 ft³**

TEMPERATURE CONVERSION PROBLEMS

These problems are a good refresher for the student or operator. Temperature is important in many plant processes. During sedimentation, the higher the temperature of the water, the faster the suspended particles, floc, and other fine materials will settle. Cold temperature water is denser than warmer water and thus particles take longer to settle. Microorganisms grow faster in warmer water than in cold water, which affects processes such as oxidation ditches, ponds, and digesters.

20. Convert 210 degrees Fahrenheit to degrees Celsius.

The equation for determining Celsius is: $°C = 5°C/9°F(°F − 32°F)$

Substitution: $°C = 5°C/9°F(210 − 32) = 5/9(178) = 98.89°C$, round to **99°C**

21. Convert 36 degrees Fahrenheit to degrees Celsius.

Equation: $C = 5°C/9°F(°F − 32°F)$

Substitution: $°C = 5°C/9°F(36 − 32) = 5/9(4) = 2.22°C$, round to **2.2°C**

22. Convert −197 degrees Celsius to degrees Fahrenheit.

The equation for determining Fahrenheit is: $°F = 9°F/5°C(°C) + 32°F$ **or use** $F = 1.8°C + 32°F$

Substitution: $°F = 9°F/5°C(−197°C) + 32 = −322.6°F$, round to **−323°F**

RATIO CALCULATIONS

Ratios are a quick and easy way to solve simple problems when a particular relationship of two variables is known and one of those variables is changed to a known value. The question now is what happens to the variable that was not changed? The final result for this unknown variable can be calculated by setting up a ratio—a "relationship" between the known variables that can be set to equal the new "relationship" with the unknown. Then, using simple algebra, solve for the unknown. The following problems are examples of ratio problems that wastewater operators may find useful in their work.

23. **A chemical pump discharges 27 mL of polymer at a speed setting of 29% and a stroke setting of 10%. If the polymer requires an increase from 27 mL to 45 mL, what should the pump speed setting be moved to? Assume pump has a linear output.**

This problem can be solved using a ratio, as follows:

$$\frac{\text{Speed setting}_1, \text{Percent}}{\text{Polymer dosage}_1, \text{mL}} = \frac{\text{Speed setting}_2, \text{Percent}}{\text{Polymer dosage}_2, \text{mL}}$$

Substitute values and solve.

$$\frac{29\%}{27 \text{ mL}} = \frac{\text{Speed setting}_2, \text{Percent}}{45 \text{ mL}}$$

$$\text{Speed setting}_2, \text{Percent} = \frac{(29\%)(45 \text{ mL})}{27 \text{ mL}} = 48.33\%, \text{round to } \mathbf{48\%}$$

24. A wastewater treatment plant has a chlorine dosage of 9.28 mg/L at a flow of 2.65 mgd. If the flow increases to 3.31 mgd, what should the chlorine dosage be increased to assuming all other parameters remain the same?

Set up a ratio.

$$\frac{\text{Chlorine dosage}_1, \text{mg/L}}{\text{Flow}_1, \text{mgd}} = \frac{\text{Chlorine dosage}_2, \text{mg/L}}{\text{Flow}_2, \text{mgd}}$$

Rearrange the ratio to solve for chlorine dosage$_2$.

$$\text{Chlorine dosage}_2, \text{mg/L} = \frac{(\text{Chlorine dosage}_1, \text{mg/L})\,(\text{Flow}_2, \text{mgd})}{\text{Flow}_1, \text{mgd}}$$

Substitute values and solve.

$$\text{Chlorine dosage}_2, \text{mg/L} = \frac{(9.28\ \text{mg/L})\,(3.31\ \text{mgd})}{2.65\ \text{mgd}} = \textbf{11.59 mg/L New chlorine dosage}$$

PERCENT AND PPM CONVERSION PROBLEMS

These problems are a good refresher for the student or operator. They are important because they are used in some dosage and mixture problems. Note: mg/L = ppm.

25. A solution was found to be 4.8% soda ash. What is the mg/L soda ash in the solution?

Know: A 1% solution = 10,000 mg/L

Thus: $\dfrac{(4.8\%)\,(10,000\ \text{mg/L})}{1\%} = \textbf{48,000 mg/L Soda ash}$

26. Convert a solution that has 26,588 ppm to percent.

Know: A 1% solution = 10,000 ppm

Thus: $\dfrac{26,588\ \text{ppm}}{10,000\ \text{ppm/1\%}} = \textbf{2.6588\% Solution}$

27. Convert 67,114 ppm to percent.

Know: A 1% solution = 10,000 ppm

Thus: $\dfrac{67,114 \text{ ppm}}{10,000 \text{ ppm}/1\%}$ = **6.7114% Solution**

PERCENT CALCULATIONS

Percent calculations are used throughout this book and are thus essential to understand.

28. If 293 is 100%, what is 43.4%.

Let x = the unknown number, that is, 43.4% of 293

$x = (293)(43.4\%/100\%) = 127.162$, round to **127**

Another way to solve these types of problems is by writing a ratio.

$293/100\% = x/43.4\%$

$x = \dfrac{(43.4\%)(293)}{100\%} = 127.162$, round to **127**

An easier way to solve these types of problems is to know that the number for 43.4% must be smaller than 293. Thus, if we multiply by the decimal for 43.4% (43.4%/100% = 0.434), we get the same answer. Simply remember that when you divide by 100%, it is the same as moving the decimal point to the left two places. If you can do this step in your head, these problems can be solved faster.

$x = (293)(0.434) = 127.162$, round to **127**

29. **Calculate the percent efficiency in removing biochemical oxygen demand (BOD_5) for a primary clarifier, if the influent BOD_5 is 228 mg/L and the effluent BOD_5 is 133 mg/L.**

Equation:

$$\text{Percent BOD}_5 \text{ removal} = \frac{(\text{Influent BOD}_5 - \text{Effluent BOD}_5)\,(100\%)}{\text{Influent BOD}_5} \text{ or } \frac{(\text{In} - \text{Out})\,(100\%)}{\text{In}}$$

$$\text{Percent BOD}_5 \text{ removal} = \frac{(228\,\text{mg/L} - 133\,\text{mg/L})\,(100\%)}{228\,\text{mg/L}} = \frac{(95\,\text{mg/L})\,(100\%)}{228\,\text{mg/L}}$$

Remember, always perform addition or subtraction within parenthesis before multiplication or division outside of the parenthesis.

Percent BOD_5 removal = (0.4167)(100%) = 41.67 %, round to **41.7% removal of BOD$_5$**

30. **If 85 is 16.2%, what is 90%? Assume two significant figures.**

Write a ratio and solve for the unknown number, x.

85/16.2% = x/90%

x = (90%)(85)/16.2% = 472.222, round to **470**

PERCENT STRENGTH BY WEIGHT SOLUTION PROBLEMS

The strength of solution calculations are important to determine so that operators can properly mix chemicals in the percentages they need for dosing a particular wastewater process or other application.

31. **If 40.1 lb of magnesium hydroxide [Mg(OH)$_2$] is dissolved in 98.4 gallons of water, what is the percent strength by weight of the Mg(OH)$_2$ solution?**

First, convert the number of gallons of water to pounds.

Number of lb = (98.4 gal)(8.34 lb/gal) = 820.656 lb of water

Next, find the percent strength of the solution.

Equation: **Percent strength** $= \dfrac{(\text{Number of lb of chemical})(100\%)}{\text{Number of lb, Water} + \text{lb, Chemical}}$

Substitute values and solve.

Percent strength $= \dfrac{[40.1\,\text{lb Mg(OH)}_2](100\%)}{820.656\,\text{lb, Water} + 40.1\,\text{lb Mg(OH)}_2}$

Percent strength $= \dfrac{[40.1\,\text{lb Mg(OH)}_2](100\%)}{860.756\,\text{lb}} =$ **4.66% Mg(OH)$_2$ solution by weight**

32. **If 15 grams (g) of lime are dissolved in 2.500 liters (L) of water, what is the percent strength by weight of the lime solution?**

Know: 1 liter = 3,785 mL = 3,785 grams

Equation: **Percent strength** $= \dfrac{(\text{Number of g of chemical})(100\%)}{\text{Number of g, Water} + \text{g, Chemical}}$

Substitute values and solve.

Percent strength $= \dfrac{(15\,\text{g Lime})(100\%)}{(2.5\,\text{L, Water})(1,000\,\text{g/L}) + 15\,\text{g Lime}}$

Percent strength $= \dfrac{(15\,\text{g Lime})(100\%)}{2,515\,\text{g}} =$ **0.60% Lime solution by weight**

33. **What is the percent strength by weight of a polymer solution, if 7.57 grams (g) of polymer is dissolved in 1.000 liters (L) of water?**

Know: 1 liter = 3,785 mL = 3,785 grams

Equation: **Percent strength** $= \dfrac{(\text{Number of g of chemical})(100\%)}{\text{Number of g, Water} + \text{g, Chemical}}$

Substitute values and solve.

Percent strength $= \dfrac{(7.57 \text{ g Polymer})(100\%)}{(1.000 \text{ L, Water})(1,000 \text{ g/L}) + 7.57 \text{ g Polymer}}$

Percent strength $= \dfrac{(7.57 \text{ g Polymer})(100\%)}{1,007.57 \text{ g}} =$ **0.75% Polymer solution by weight**

PERCENT SOLIDS BY WEIGHT CALCULATIONS

Operators use percent solids calculations to determine efficiency of different unit processes, as well as to determine how much waste will require disposal.

34. **What are the percent total solids by weight in a sludge sample that weighed 295 grams before drying and 16.3 grams after drying?**

Equation: **Percent total solids** $= \dfrac{(\text{Dry sample in grams})(100\%)}{\text{Sludge sample in grams}}$

Percent total solids $= \dfrac{(16.3 \text{ grams})(100\%)}{295 \text{ grams}} =$ **5.53% Total solids by weight**

35. **If the percent total solids by weight are 4.72% and the dried sample (total solids) weighed 9.84 grams, what must have been the weight of the sludge sample before it was dried?**

Equation: **Percent total solids** $= \dfrac{(\text{Dry sample in grams})(100\%)}{\text{Sludge sample in grams}}$

Rearrange the equation.

$$\text{Sludge sample, grams} = \frac{(\text{Dry sample in grams})(100\%)}{\text{Percent total solids}}$$

Substitute values and solve.

$$\text{Sludge sample, grams} = \frac{(9.84\text{ grams})(100\%)}{4.72\%} = 208.47\text{ grams, round to }\textbf{208 grams}$$

PERCENT VOLATILE SOLIDS REDUCTION

The percent volatile solids reduction calculations indicate the effectiveness of the digested sludge process when compared to the volatile solids in the influent. The higher the percent volatile solids reduced or destroyed, the more stable the organic matter in the digester becomes and the more gas that is produced.

36. **Calculate the percent volatile solids (VS) reduction by weight, if the digester influent sludge has a VS content of 65% and the digester effluent sludge has a VS content of 51%.**

First, convert percentage to decimal form by dividing by 100%.

65%/100% = 0.65

51%/100% = 0.51

Equation: **Percent VS reduction** =

$$\frac{(\text{Percent influent VS} - \text{Percent effluent VS})(100\%)}{[\text{Percent influent VS} - (\text{Percent influent VS})(\text{Percent effluent VS})]}$$

$$\text{Percent VS reduction} = \frac{(0.65 - 0.51)(100\%)}{0.65 - (0.65)(0.51)} = \frac{0.14(100\%)}{0.65 - 0.3315}$$

$$\text{Percent VS reduction} = \frac{14\%}{0.3185} = 43.96\%\text{, round to }\textbf{44\% VS reduction by weight}$$

37. If the digester influent sludge has a volatile solids (VS) content of 61% and the digester effluent sludge has a VS content of 48.5% by weight, calculate the percent VS reduction.

First, convert percentage to decimal form by dividing by 100%.

$61\%/100\% = 0.61$

$48.5\%/100\% = 0.485$

Equation: **Percent VS reduction** $= \dfrac{(\text{Influent} - \text{Effluent})(100\%)}{\text{Effluent} - (\text{Effluent})(\text{Influent})}$

Percent VS reduction $= \dfrac{(0.61 - 0.485)(100\%)}{0.61 - (0.61)(0.485)} = \dfrac{0.125(100\%)}{0.61 - 0.29585}$

Percent VS reduction $= \dfrac{12.5\%}{0.31415} = 39.79\%$, round to **40% VS reduction by weight**

PERCENT MOISTURE REDUCTION PROBLEMS

This calculation will tell the operator the efficiency of the moisture reduction process.

38. What is the percent moisture reduction by weight for a digester, if the raw biosolids is 8.1% solids and the digested biosolids is 15.9%?

Equation: **Percent moisture reduction** $=$

$$\frac{(\text{Percent influent moisture} - \text{Percent moisture, after digestion})(100\%)}{[\text{Percent influent moisture} - (\text{Percent influent moisture})(\text{Percent moisture, after digestion})]}$$

First, convert the percentages for solids to moisture percent then to decimal form for easier substitution.

Raw biosolids $= 100\% - 8.1\% = 91.9\%/100\% = 0.919$

Digested biosolids $= 100\% - 15.9\% = 84.1\%/100\% = 0.841$

Substitute values and solve.

$$\text{Percent moisture reduction} = \frac{(0.919 - 0.841)(100\%)}{[0.919 - (0.919)(0.841)]}$$

Simplify:

$$\text{Percent moisture reduction} = \frac{(0.078)(100\%)}{(0.919 - 0.772879)}$$

$$\text{Percent moisture reduction} = \frac{(0.078)(100\%)}{0.146121} \quad \textbf{53\% Moisture reduction by weight}$$

39. What is the percent moisture reduction by weight for a digester, if the raw biosolids is 6.8% solids and the digested biosolids is 13.9%?

Equation: **Percent moisture reduction** $=$

$$\frac{(\text{Percent influent moisture} - \text{Percent moisture, after digestion})(100\%)}{[\text{Percent influent moisture} - (\text{Percent influent moisture})(\text{Percent moisture, after digestion})]}$$

First, convert the percentages for solids to moisture percent then to decimal form for easier substitution.

Raw biosolids $= 100\% - 6.8\% = 93.2\%/100\% = 0.932$

Digested biosolids $= 100\% - 13.9\% = 86.1\%/100\% = 0.861$

Substitute values and solve.

$$\text{Percent moisture reduction} = \frac{(0.932 - 0.861)(100\%)}{[0.932 - (0.932)(0.861)]}$$

Simplify:

$$\text{Percent moisture reduction} = \frac{(0.071)(100\%)}{(0.932 - 0.802452)}$$

$$\text{Percent moisture reduction} = \frac{(0.071)(100\%)}{0.129548} = \textbf{55\% Moisture reduction by weight}$$

VOLATILE SOLIDS PUMPING CALCULATIONS

These calculations are used as a planning tool by the operator. By knowing the pumping rate of volatile solids into a digester, an operator can make sure it is not overloaded, which would adversely affect the digester's operation and performance. See Figures 2, 4, 5, and 6 in Appendix E for four types of wastewater plants using a digester.

40. **How many lb/day of volatile solids (VS) are sent to a digester, if 56% by weight of the 2,344 lb/day of solids sent to the digester are volatile?**

Equation: **VS, lb/day = (Number of lb/day, sent to digester)(Percent VS)/100%**

VS, lb/day = (2,344 lb/day Solids)(56% VS)/100% = 1,312.64, round to **1,300 lb/day of VS**

41. **Given the following data, how many lb/day of volatile solids (VS) are pumped to a digester?**

Pumping rate = 7,745 gpd
Solids content = 6.25%
Volatile solids = 59.5%

Equation: **VS, lb/day =**

$$\textbf{(Number of gpd to digester)} \frac{\textbf{(Percent solids)}}{\textbf{100\%}} \ \frac{\textbf{(Percent VS)}}{\textbf{100\%}} \textbf{(8.34 lb/gal)}$$

VS, lb/day = (7,745 lb/day Solids)$\frac{(6.25\%)}{100\%}$ $\frac{(59.5\% \text{ VS})}{100\%}$(8.34 lb/gal)

VS, lb/day = 2,402.06 lb/day, round to **2,400 lb/day of VS**

42. **Given the following data, how many lb/day of volatile solids (VS) are pumped to a digester?**

Pumping rate = 4,055 gpd
Volatile solids = 61%
Solids content = 7.1%

Equation:

$$\text{VS, lb/day} = (\text{Number of gpd to digester})\frac{(\text{Percent solids})}{100\%}\frac{(\text{Percent VS})}{100\%}(8.34 \text{ lb/gal})$$

$$\text{VS, lb/day} = (4,055 \text{ lb/day solids})\frac{(7.1\%)}{100\%}\frac{(61\% \text{ VS})}{100\%}(8.34 \text{ lb/gal})$$

VS, lb/day = 1,464.69, round to **1,500 lb/day of VS**

CALCULATIONS FOR ARITHMETIC MEAN, MEDIAN, RANGE, MODE, AND GEOMETRIC MEAN

These calculations are good tools for planning and evaluating plant processes.

43. **What is the average mg/L per day of volatile acids coming off a digester given the following data?**

Mon.	Tues.	Wed.	Thurs.	Fri.	Sat.	Sun.
234	261	280	272	259	257	244

Equation: **Avg. volatile acids, mg/L/day** $= \dfrac{\text{Sum of volatile acids, mg/L/day}}{\text{Total time, days}}$

Avg. volatile acids, mg/L/day $= \dfrac{234 + 261 + 280 + 272 + 259 + 257 + 244}{7 \text{ days}}$

Avg. volatile acids, mg/L/day = 258.14 mg/L/day, round to **258 mg/L/day of Volatile acids**

44. **What is the average number of gallons per day (gpd) per square foot of nitrogen removal given the following data?**

Mon.	Tues.	Wed.	Thurs.	Fri.	Sat.	Sun.
1.53	1.59	1.60	1.62	1.64	1.68	1.71

Equation: **Avg. nitrogen removal, gpd/ft²** $= \dfrac{\text{Sum of nitrogen removal, gpd/ft}^2}{\text{Total time, days}}$

Avg. nitrogen removal, gpd/ft² $= \dfrac{1.53 + 1.59 + 1.60 + 1.62 + 1.64 + 1.68 + 1.71}{7 \text{ days}}$

Avg. nitrogen removal, gpd/ft² $=$ **1.62 gpd/ft² of Nitrogen removal**

45. **Calculate the moving (running) average for BOD$_5$ removal during days 8, 9, and 10 given the following data:**

1—212 mg/L	9—226 mg/L
2—231 mg/L	10—211 mg/L
3—244 mg/L	11—245 mg/L
4—235 mg/L	12—206 mg/L
5—217 mg/L	13—193 mg/L
6—202 mg/L	14—188 mg/L
7—194 mg/L	15—189 mg/L
8—209 mg/L	16—204 mg/L

Day 8, 7-day moving average:

Add day 8 BOD$_5$ to the previous 6 days.

7-day average $= \dfrac{231 + 244 + 235 + 217 + 202 + 194 + 209}{7 \text{ days}} =$ **219 mg/L BOD$_5$**

Day 9, 7-day moving average:

Add day 9 BOD$_5$ to the previous 6 days.

$$\text{7-day average} = \frac{244 + 235 + 217 + 202 + 194 + 209 + 226}{7 \text{ days}} = \textbf{218 mg/L BOD}_5$$

Day 10, 7-day moving average:

Add day 10 BOD$_5$ to the previous 6 days.

$$\text{7-day average} = \frac{235 + 217 + 202 + 194 + 209 + 226 + 211}{7 \text{ days}} = \textbf{213 mg/L BOD}_5$$

46. Given the following data, calculate the unknowns.

Note: A scientific calculator is required for determining the geometric mean.

Day	Effluent BOD$_5$, mg/L	Unknown
Monday	28	a. Arithmetic mean, mg/L
Tuesday	32	b. Median, mg/L
Wednesday	34	c. Range, mg/L
Thursday	32	d. Mode, mg/L
Friday	29	e. Geometric mean, mg/L
Saturday	23	
Sunday	35	

a. Calculate the arithmetic mean of BOD$_5$, mg/L and use the exact rule for rounding.

Equation: **Arithmetic mean** $= \dfrac{\text{Sum of all measurements}}{\text{Number of measurements}}$

Arithmetic Mean $= \dfrac{28 + 32 + 34 + 32 + 29 + 23 + 35}{7} = \textbf{30.4 mg/L BOD}_5$

b. Determine the median of BOD$_5$ mg/L.

To determine the median, put the chlorine dosages in ascending order and choose the middle value.

1	2	3	4	5	6	7
23	28	29	**32**	32	34	35

In this case the middle value is **32 mg/L BOD$_5$**

c. Determine the range of BOD$_5$ mg/L.

Equation: **Range = Largest value − Smallest value**

Range, mg/L = 35 mg/L − 23 mg/L = **12 mg/L BOD$_5$**

d. Determine the mode of BOD$_5$ mg/L.

Mode is the measurement that occurs most frequently.

In this case it is **32 mg/L BOD$_5$**

e. Calculate the geometric mean and use the exact rule for rounding.

Equation: **Geometric mean $= [(x_1)(x_2)(x_3)(x_4).....(x_n)]^{1/n}$**

Geometric mean, mg/L $= [(23)(28)(29)(32)(32)(34)(35)]^{1/7}$

Geometric mean, mg/L $= (22757826560)^{1/7}$

Geometric mean, mg/L = **30.2 mg/L BOD$_5$**

AREA PROBLEMS

Areas are important to determine for a number of reasons including knowing the "footprint" of a tank or pond or the area of a particular process to make further calculations in other wastewater problems.

47. What is the area of an aeration tank that is 85 ft long and 9.5 ft wide?

Equation for a rectangular area is: **Area = (Length)(Width)**

Area, ft^2 = (85 ft)(9.5 ft) = 807.5 ft^2, round to **810 ft^2**

48. What is the area of a trench that is 126 ft by 3.3 ft?

Equation: **Area = (Length)(Width)**

Area, ft^2 = (126 ft)(3.3 ft) = 415.8 ft^2, round to **420 ft^2**

49. What is the area of a digester that is 59.9 ft in diameter?

Equation: **Area = πr^2,** where π = 3.14

First, find the radius.

Radius = Diameter/2 = 59.9/2 = 29.95

Area of tank, ft^2 = (3.14)(29.95 ft)(29.95 ft) = 2,816.59 ft^2, round to **2,820 ft^2**

50. What is the area of a tank, if the radius is 49.8 ft?

Equation: **Area $= \pi r^2$ or $(0.785)(\text{Diameter})^2$**

In this problem let us use Area $= \pi r^2$ since the radius is already given.

Area, ft^2 $= (3.14)(49.8 \text{ ft})(49.8 \text{ ft}) = 7787.33$ ft^2, round to **7790 ft^2**

51. What is the area of a circular reservoir, if it has a diameter of 2,170 ft?

Equation: **Area $= (0.785)(\text{Diameter})^2$**

Area of tank, ft^2 $= (0.785)(2,170 \text{ ft})(2,170 \text{ ft}) = 3696486.5$ ft^2, round to **3,700,000 ft^2**

52. If the surface area of a stabilization pond is 12,805 ft^2 and one side measures 50.0 ft, what is the length of the other side?

Equation: **Area $= (\text{Length})(\text{Width})$**

Length, ft $= \dfrac{\text{Area of } 12,805 \text{ ft}^2}{50.0 \text{ ft}} = 256.1$ ft, round to **256 ft**

CIRCUMFERENCE PROBLEMS

The circumference is important to know for calculating the area of a circular tank or the area of a particular process to make further calculations in a problem. For example, an operator needs to calculate the weir overflow rate on a circular clarifier: given the diameter, the circumference or length of the weir can be calculated, and thus the weir overflow rate can be determined from this.

53. What is the circumference of a tank that is 75 ft in diameter as measured to the nearest foot?

Equation: **Circumference $= \pi(\text{Diameter})$**

Circumference $= (3.14)(75 \text{ ft}) = 235.5$ ft, round to **236 ft**

54. What is the circumference of a clarifier, if the radius is 40.0 ft?

Equation: **Circumference = 2π(radius) or 2πr**

Circumference = 2(3.14)(40.0 ft) = 251.2 ft, round to **251 ft**

VOLUME PROBLEMS

Volumes are very important to determine because many problems in the wastewater field require the volume to be known before the rest of the calculations can be made. Knowing the volume of a particular process can also help the operator plan and make proper decisions in the treatment of wastewater.

55. A circular tank has a radius of 29.85 ft and is 12.1 ft high at the spill point. What is the capacity of the tank in cubic feet?

The volume equation for a circular tank is: **Volume = πr²(Height)**, where r is the radius.

Volume of tank in ft³ = 3.14(29.85 ft)(29.85 ft)(12.1 ft) = 33,853.5 ft³, round to **33,900 ft³**

56. If a stabilization pond is 360 ft long, 64.5 ft wide, and 7.8 ft deep, what is the number of cubic feet in a stabilization pond?

The volume equation for a rectangular basin is: **Volume = (Length)(Width)(Depth)**

Volume = (L)(W)(D) = (360 ft)(64.5 ft)(7.8 ft) = 181,116 ft³, round to **180,000 ft³**

57. A circular tank has a radius of 24.0 ft and is 21.2 ft high at the spill point. What is the capacity of the tank in cubic feet?

The volume equation for a circular tank is: **Volume = πr²(Height)**, where r is the radius

Volume of tank in ft³ = 3.14(24.0 ft)(24.0 ft)(21.2 ft) = 38,343 ft³, round to **38,300 ft³**

58. Calculate the volume in cubic feet for a pipeline that is 12.0 inches in diameter and 1,080 ft long.

First, convert the diameter to feet.

$$(12.0 \text{ in.})\frac{(1 \text{ ft})}{12 \text{ in.}} = 1.00 \text{ ft (Diameter)}$$

Then, convert the diameter to the radius.

radius = Diameter /2 = 1.00 ft/2 = 0.500 ft (radius)

Equation for the volume of a pipe in cubic feet is: πr^2**(Length)** or **(0.785 (Diameter)^2(Length)**

Using the first equation, the Volume, ft^3 = (3.14)(0.500 ft)(0.500 ft)(1,080 ft)

Volume, ft^3 = 847.8 ft^3, round to **848 ft^3**

59. How many gallons would be in the pipe for the problem above?

(847.8 ft^3)(7.48 gal/ft^3) = 6,341.5 gal, round to **6,340 gal**

60. What is the volume of a pipeline that is 14 inches in diameter and 455 ft long?

Equation: **Volume, ft^3 = (0.785)(Diameter)^2(Length)**

First, convert the diameter to feet.

$$(14.0 \text{ in.})\frac{(1 \text{ ft})}{12 \text{ in.}} = 1.167 \text{ ft (Diameter)}$$

Volume, ft^3 = (0.785)(1.167 ft)(1.167 ft)(455 ft) = 486.4 ft^3, round to **490 ft^3**

61. What is the volume of a conical tank in cubic feet that has a radius of 10.0 ft and a height of 17 ft?

Equation: **Volume, ft³ = 1/3πr²(Height or Depth)**

Substitute values and solve.

Volume, ft³ = 1/3π(10.0 ft)(10.0 ft)(17 ft) = 1,779 ft³, round to **1,800 ft³**

62. Determine the volume in cubic feet for a pipeline that is 18 inches in diameter and 552 ft long.

First, convert the diameter to feet.

$(18 \text{ in.})\dfrac{(1 \text{ ft})}{12 \text{ in.}}$ = 1.5 ft in Diameter

Equation: **Volume, ft³ = (0.785)(Diameter)²(Length)**

Volume, ft³ = (0.785)(1.5 ft)(1.5 ft)(552 ft) = 974.97 ft³, round to **970 ft³**

63. A dry chemical feed tank is conical at the bottom and cylindrical at the top. If the diameter of the cylinder is 20.0 ft with a depth of 36.2 ft and the cone depth is 10.1 ft, what is the approximate volume of the tank in cubic feet and gallons?

First, find the volume of the cone in cubic feet.

Volume, ft³ = 1/3πr²(Depth)

Where the radius = Diameter/2 = 20.0 ft/2 = 10.0 ft

Volume, ft³ = 1/3(3.14)(10.0 ft)(10.0 ft)(10.1 ft) = 1,057.13 ft³

Next, find the volume of the cylindrical part of the tank.

Volume, ft^3 = πr^2(Depth) = (3.14)(10.0 ft)(10.0 ft)(36.2 ft) = 11,366.8 ft^3

Then, add the two volumes for the answer.

Total volume, ft^3 = 1,057.13 ft^3 + 11,366.8 ft^3 = 12,423.93 ft^3, round to **12,400 ft^3**

To find the number of gallons multiply the total number of cubic feet by 7.48 gal/ft^3.

Number of gallons = (12,423.93 ft^3)(7.48 gal/ft^3) = 92,930.996 gal, round to **92,900 gal**

64. What is the water volume in gallons of a trapezoidal channel given the following data:

Length = 365 ft
Water width at surface = 11.5 ft
Depth = 6.3 ft
Water width at bottom = 5.8 ft

Equation for a trapezoidal channel is: **Volume, gal** $= \dfrac{(b_1 + b_2)}{2}$**(Depth)(Length)(7.48 gal/ft^3)**

Where b$_1$ = Water width at surface, b$_2$ = Water width at bottom, and altitude = depth

Volume, gal $= \dfrac{(11.5 \text{ ft} + 5.8 \text{ ft})}{2}$(6.3 ft)(365 ft)(7.48 gal/ft^3) $= \dfrac{(17.3)}{2}$(6.3 ft)(365 ft)(7.48 gal/ft^3)

Volume, gal = 148,782.25 gal, round to **150,000 gal**

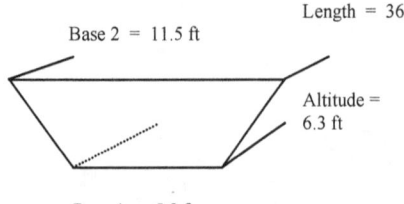

Length = 365 ft

Base 2 = 11.5 ft

Altitude = 6.3 ft

Base 1 = 5.8 ft

Trapezoid – a four sided plane figure that has two parallel sides and two non-parallel sides.

DENSITY CALCULATIONS

The density of a substance is the amount of mass for a given volume. It is usually expressed as lb/gal or lb/ft^3 in the English system or as g/cm^3, kg/L, or kg/m^3 in the metric system. Mass is defined as the quantity of matter as determined from Newton's second law of motion or by its weight. Weight is defined as the force that gravitation exerts upon a body and is equal to the mass of the body times the local acceleration of gravity.

65. A substance weighs 771 grams and occupies a space of 300.0 cubic centimeters (cm^3). What is its density in grams/cm^3?

Equation: **Density = Mass/Volume**

Density = 771 grams/300.0 cm^3 = **2.57 grams/cm^3**

66. What is the density in lb/gal of a 1.000-liter solution that weighs 2.85 lb?

First, convert mL to gallons.

(1.000 liter)(1 gal/3.785 liters) = 0.2642 gal

Equation: **Density = Mass/Volume**

Density of solution = 2.85 lb/0.2642 gal = **10.8 lb/gal**

67. **The density of an unknown substance is 4.55 grams/cm³. How much space would this substance occupy in cm³, if it weighed 8.09 lb?**

First, convert the number of lb to grams (g).

Number of g = (Number of lb)(454 g/1 lb)

Substitution: Number of g = (8.09 lb)(454 g/1 lb) = 3,672.86 g

We know that 4.55 grams of the substance occupies 1 cm³ by knowing its density. To get the space 3,672.86 grams occupies we only need to divide by the density.

$$\text{Space occupied by substance} = \frac{3{,}672.86 \text{ g}}{4.55 \text{ g/cm}^3} = 807.22 \text{ cm}^3, \text{ round to } \mathbf{807 \text{ cm}^3}$$

68. **If a substance weighs 10.38 lb/gal, what is the density of a solution in g/cm³?**

Equation: **Number of g/cm³ = (Number of lb/gal)(454 g/1 lb)(1 gal/3,785 cm³)**

Number of g/cm³ = (10.38 lb/gal)(454 g/1 lb)(1 gal/3,785 cm³)

Number of g/cm³ = 1.245 g/cm³, round to **1.24 g/cm³**

69. **The density of an unknown substance is 2.09 grams/cm³. How much space would this substance occupy in cm³, if it weighed 4.4 lb?**

First, convert the number of lb to grams (g).

Number of g = (Number of lb)(454 g/1 lb)

Substitution: Number of g = (4.4 lb)(454 g/1 lb) = 1,997.6 g

We know that 2.09 grams of the substance occupies 1 cm³ by knowing its density. To get the space 1,997.6 grams occupies, we only need to divide by the density.

Space occupied by substance $= \dfrac{1,997.6\,\text{g}}{2.09\,\text{g/cm}^3} = 955.79$ cm³, round to **960 cm³**

SPECIFIC GRAVITY OF LIQUIDS

Specific gravity compares the density of one substance to another. Water is the standard for liquids and is equal to 1.

70. The specific gravity (sp gr) of a liquid solution is 1.30. How many lb will 1 ft³ weigh?

Know: 62.4 lb = 1 ft³ for water

Equation: **lb/ft³ = (sp gr)(62.4 lb/ft³)**

Substitute values and solve.

lb/gal of liquid solution = (1.30)(62.4 lb/ft³) = **81.1 lb for 1 ft³ of the liquid solution**

71. The density of an unknown liquid is 65.8 lb/ft³. What is the specific gravity of the liquid?

Know: Water has a density of 62.4 lb/ft³. Divide the density of the unknown by the density of water.

Equation: **Specific gravity (sp gr) = Density of substance/Density of water**

Sp gr of Unknown Substance $= \dfrac{65.8\,\text{lb/ft}^3}{62.4\,\text{lb/ft}^3} = $ **1.05 sp gr**

72. What is the specific gravity for a solution that weighs 9.07 lb/gal?

Know: Density of water can also be expressed as lb/gal, or 8.34 lb/gal.

$$\text{Sp gr} = \frac{9.07\ \text{lb/gal}}{8.34\ \text{lb/gal}} = 1.088\ \text{sp gr, round to } \mathbf{1.09\ sp\ gr}$$

SPECIFIC GRAVITY OF SOLIDS

Specific gravity (sp gr) compares the density of one substance to another. Water is the standard for solids and is equal to 1.

73. A piece of metal that weighs 35.4 kilograms in air is weighed in water and found to be 22.8 kilograms. What is the specific gravity of this metal?

First, subtract the weight in air from the weight in water to determine the loss of weight in water.

Number of kilograms = 35.4 kg − 22.8 kg = 12.6 kg is weight loss in water

Next, find the specific gravity by dividing the weight of the metal in air by the weight loss in water.

Sp gr = 35.4 kg/12.6 kg = 2.8095, round to **2.81 sp gr**

74. A piece of metal that weighs 18.1 kilograms in air is weighed in water and found to be 12.2 kilograms. What is the specific gravity of this metal?

First, subtract the weight in air from the weight in water to determine the loss of weight in water.

Number of kilograms = 18.1 kg − 12.2 kg = 5.9 kg is weight loss in water

Next, find the specific gravity by dividing the weight of the metal in air by the weight loss in water.

Sp gr = 18.1 kg/5.9 kg = 3.0678, round to **3.07 sp gr**

PRESSURE PROBLEMS

Pressure is the measure of force against a surface and is usually expressed as force per unit area. In the English system the units are usually in lb/in.² or lb/ft². Scientists and engineers usually use the metric system, where pressure is measured in Pascals (Pa). One Pascal is equal to a force of 1 Newton per square meter. A Newton is equal to the force required to accelerate 1 kilogram one meter per second per second (1 kg·m/s²). You can also have kilopascals (kPa), megapascals (mPa), and gigapascals (gPa). Also: 1 Pascal = 10 dyne/cm² = 0.01 mbar. 1 atm = 101,325 Pascals = 760 mm Hg = 760 torr (Torricelli barometer) = 14.7 psi. *Note:* psi = pounds per square inch.

75. What is the height of water in a storage tank if the psi is 6.80 at the bottom?

Equation: **Height, ft = (psi)(2.31 ft/psi)**

Substitute values and solve.

Height, ft = (6.80 psi)(2.31 ft/psi) = **15.7 ft, Height**

76. What is the psi at the bottom of a tank, if the water level is 15.52 ft deep?

Equation: $\textbf{psi} = \dfrac{\text{Depth, ft}}{2.31 \text{ ft/psi}}$

$\text{psi} = \dfrac{15.52 \text{ ft}}{2.31 \text{ ft/psi}} = \textbf{6.72 psi}$

77. What is the depth of water in a pond, if the psi is 4.88?

Equation: **Depth, ft = (psi)(2.31 ft/psi)**

Depth, ft = (4.88 psi)(2.31 ft/psi) = **11.3 ft deep**

78. **If the height of water in a tank is 11.08 ft, what is the pressure at the bottom of the tank in lb/ft²?**

Equation using pressure in lb/ft² is:

Pressure, lb/ft² = (Height or Depth, ft)(Density, 62.4 lb/ft³)

Substitute values and solve.

Pressure, lb/ft² = (11.08 ft)(62.4 lb/ft³) = 691.39 lb/ft², round to **691 lb/ft²**

HYDRAULIC PRESS CALCULATIONS

Hydraulic press calculations have two fundamental principles: total force equals pressure applied times area the pressure is applied to, and the force applied to a liquid will be equally distributed within that liquid.

79. **A force of 78.5 pounds is applied to a small cylinder on a hydraulic jack. The diameter of the small cylinder is 10.0 inches. If the diameter of the large cylinder is 2.25 ft, what is the total lifting force?**

Equation: $\textbf{Pressure} = \dfrac{\text{Total force, lb}}{\text{Area, ft}^2}$ for pressure on the small cylinder

First, convert 10.0 inches to feet.

(10.0 in.)(1 ft/12 in.) = 0.833 ft

Substitution:

$\text{Pressure} = \dfrac{78.5\ \text{lb}}{(0.785)(0.833\ \text{ft})(0.833\ \text{ft})} = 144.12\ \text{lb/ft}^2$

Next, calculate the total force on the large cylinder.

Equation: **Total force = (Pressure)(Area), where area = (0.785)(Diameter)²**

Total Force, lb = (144.12 lb/ft²)(0.785)(2.25 ft)(2.25 ft) = 572.74 lb, round to **573 lb**

80. **A small cylinder on a hydraulic jack is 8.0 inches in diameter. A force of 245 pounds is applied to the small cylinder. If the diameter of the large cylinder is 2.50 ft, what is the total lifting force?**

Equation: **Pressure** $= \dfrac{\text{Total Force, lb}}{\text{Area, ft}^2}$ for pressure on the small cylinder

First, convert 8.0 inches to feet.

(8.0 in.)(1 ft/12 in.) = 0.667 ft

Pressure, lb/ft² $= \dfrac{245\ \text{lb}}{(0.785)\,(0.667\ \text{ft})\,(0.667\ \text{ft})} = 701.53$ lb/ft²

Next, calculate the total force on the large cylinder.

Equation: **Total force = (Pressure)(Area)**

Total Force = (701.53 lb/ft²)(0.785)(2.50 ft)(2.50 ft) = 3,441.88 lb, round to **3,400 lb**

SCREENING MATERIAL REMOVAL CALCULATIONS

The amount of screening debris should be calculated by operators so that they can plan and properly dispose of the material. A record should be kept each time for the amount of material removed from the screening pits. Screenings are usually disposed of by landfill, incinerated, or ground and returned to the wastewater process. They are very odorous and will attract flies. See the figures in Appendix E for placement of wastewater screens.

81. **A wastewater treatment plant processes an average of 2.56 mgd. If the screenings average 89 gallons per day (gpd), what is the cubic foot of screenings per million gallons?**

First, determine the amount of cubic feet in 89 gallons.

Number of ft^3 = 89 gal/7.48 gal/ft^3 = 11.898 ft^3

Next, determine the cubic foot of screenings per million gallons.

Equation: **Screenings, ft³/mil gal** $= \dfrac{\text{Number of ft}^3/\text{day}}{\text{Number of mgd}}$

Screenings, ft³/mil gal $= \dfrac{11.898 \text{ ft}^3/\text{day}}{2.56 \text{ mgd}}$ = 4.65 ft³/mil gal, round to **4.6 ft³/mil gal**

82. **During the month of April an average of 46.5 gallons of screenings were removed each day. What were the total cubic feet of screenings removed that month?**

Total screenings, ft^3 $= \dfrac{(46.5 \text{ gal/day})(30 \text{ days})}{7.48 \text{ gal/ft}^3}$ = 186.497 ft^3, round to **186 ft³**

SCREENING PIT CAPACITY CALCULATIONS

The operator needs to know the capacity of a screening pit so he or she knows when it should be cleaned based on past records of material removed (above calculations).

83. **An average of 3.25 ft³ of screenings is added to a screening pit each day. If the pit is 6.0 ft by 10 ft and 4.2 ft deep, how many days will it take to fill the pit?**

First, determine the volume of the pit in ft³.

Pit volume, ft³ = (6.0 ft)(10 ft)(4.2 ft) = 252 ft³

Equation: **Number of days to fill** $= \dfrac{\text{Pit volume, ft}^3}{\text{Screenings removed, ft}^3/\text{day}}$

Number of days to fill $= \dfrac{252 \text{ ft}^3}{3.25 \text{ ft}^3/\text{days}} = 77.5$ days, round to **78 days**

84. **A new screening pit is being designed to replace the old one at a wastewater treatment plant. The engineer wants to design the pit to hold enough screenings for exactly 60 days. If the number of cubic feet normally removed by the plant is 76.5 gal/day in worst-case scenarios, determine the size in ft³ the pit should be.** *Note*: **round to the nearest cubic foot.**

Pit size, ft³ $= \dfrac{(76.5 \text{ gal/day})(60 \text{ days})}{7.48 \text{ gal/ft}^3} = 613.64$ ft³, round to **614 ft³**

85. **An average of 72 gallons per day of screenings is removed from a screening pit that has a capacity of 7.5 yd³. How long will it take to fill the pit in days, if the screenings are not removed?**

First, convert yd³ to ft³.

Know: 1 yd³ = 27 ft³

Number of ft³ = (7.5 yd³)(27 ft³/yd³) = 202.5 ft³

Number of days $= \dfrac{(202.5 \text{ ft}^3)(7.48 \text{ gal/ft}^3)}{72 \text{ gal/day}} = $ **21 days**

GRIT REMOVAL CALCULATIONS

Grit removal is important for the same reason as screening removal— planning for proper disposal. Grit channels are important in wastewater treatment because by removing the grit from the waste it prevents wear on pumps and deposition in pipelines or channels. It also prevents grit from accumulating in other processes such as digesters or biological contactors. Not all wastewater treatment plants have grit channels and they are not always placed after screens or comminutors. See the figures in Appendix E for where grit channels are commonly placed in different treatment plants.

86. **A wastewater plant removes 42 gallons of grit during the processing of 3.08 mil gal. What is the ft³/mil gal removal rate during this interval?**

Equation: **Grit removal, ft³/mil gal** $= \dfrac{\text{Number of gallons removed}}{(7.48 \text{ gal/ft}^3)\,(\text{mil gal treated})}$

Grit removal, ft³/mil gal $= \dfrac{42 \text{ gal}}{(7.48 \text{ gal/ft}^3)\,(3.08 \text{ mil gal})} =$ **1.8 ft³/mil gal**

87. **Calculate the number of gallons of grit removed from a wastewater plant on a day that the plant treated 2.17 mil gal and the grit removal rate during that time interval was 1.62 ft³/mil gal.**

Grit removal, gal $= (1.62 \text{ ft}^3/\text{mil gal})(2.17 \text{ mil gal})(7.48 \text{ gal/ft}^3) =$ **26.3 gal**

RECIRCULATION RATIO PROBLEMS

The recirculation ratio calculation is used to help operators keep flow variations to a trickling filter to a minimum. This ratio can also help solve process problems such as increasing the rate of hydraulic loading when needed, reducing odors and filter flies, or to prevent the trickling filter from drying out during low flows. Recirculation is from the secondary clarifier or trickling filter effluent. The ratio usually ranges from 1:1 to 2:1.

88. **What is the recirculation ratio for a trickling filter, if the influent to a trickling filter plant is 1.09 mgd and the recirculation of the trickling filter effluent is 1.83 mgd?**

Equation: $\textbf{Ratio} = \dfrac{\text{Recirculated flow}}{\text{Plant influent flow}}$

Ratio = 1.83 mgd/1.09 mgd = **1.68**

89. **What must have been the trickling filter's effluent flow in mgd, if the influent flow to the wastewater plant was 2.17 mgd and the recirculation ratio was 1.14?**

Equation: $\textbf{Ratio} = \dfrac{\text{Recirculated flow}}{\text{Plant influent flow}}$

Rearrange the equation to solve for the recirculation flow (return of the trickling filter's effluent).

Recirculated flow = (Ratio)(Plant influent flow)

Substitute values and solve.

Recirculated flow = (1.14)(2.17) = **2.47 mgd**

DETENTION TIME CALCULATIONS

Detention time is simple the time period that starts when wastewater flows into a basin or tank and ends when it flows out of the basin or tank. Detention time is usually calculated for wastewater ponds, oxidation (aerobic) ditches, and clarifiers. Detention times are theoretical, since basins begin to fill with settled sludge and other debris. This causes the true detention time to constantly change (decrease). While it is true sludge removals will cause the detention time to increase, the true detention time will always be less than theoretical. Also, flows through a basin are never perfectly laminar and thus cause a further decrease in the true detention time. See Figures 5 and 6 in Appendix E for two types of wastewater plants using ponds.

90. What is the detention time in days for a wastewater treatment pond that is 382 ft long, 159 ft wide, and 4.75 ft in depth, if the flow is 0.0562 mgd?

First, convert the number of mgd to gallons per day (gpd) in the waste treatment pond.

(0.0562 mgd)(1,000,000/1 mil) = 56,200 gpd

Next, determine the volume of the waste treatment pond.

Equation: **Volume, gal = (Length)(Width)(Depth)(7.48 gal/ft³)**

Volume, gal = (382 ft)(159 ft)(4.75 ft)(7.48 gal/ft³) = 2,158,021 gal

Write the equation with units asked for in the question.

$$\textbf{Detention time, days} = \frac{\text{Volume, gal}}{\text{Flow rate, gpd}}$$

$$\text{Detention time, days} = \frac{2,158,021\ \text{gal}}{56,200\ \text{gpd}} = \textbf{38.4 days}$$

Detention times are theoretical since basins begin to fill with solids as time passes.

91. **Find the detention time in days for an aeration pond that averages 462 ft in length, 391 ft in width, and 3.52 ft in depth, and flow through the pond is 75,000 gal per day.**

First, determine the volume in gallons for the aeration pond.

Volume, gal = (Length)(Width)(Depth)(7.48 gal/ft³)

Volume, gal = (462 ft)(391 ft)(3.52 ft)(7.48 gal/ft³) = 4,756,232 gal

Then, determine the detention time.

Equation: Detention time, days = $\dfrac{\text{Volume, gal}}{\text{Flow, gpd}}$

Detention time, days = $\dfrac{\text{4,756,232 gal}}{\text{75,000 gpd}}$ = 63.4 days

WEIR OVERFLOW RATE PROBLEMS

A weir is like a small dam, gate, notch, or other barrier placed across a basin to help regulate water out of the basin. The weir overflow rate is used to determine the velocity of wastewater over the weir. The velocity informs the operator about the efficiency of the sedimentation process. At constant wastewater flow, the shorter the length of the weir the faster the water velocity will be out of the basin. Conversely, the longer the weir length the slower the velocity will be out of the basin. See Figures 1, 2, 3, 7, and 8 in Appendix E for five types of wastewater plants using a clarifier.

92. **A rectangular clarifier has a weir length of 175 ft. What is the weir overflow rate in gpd/ft, if the flow is 0.95 mgd?**

Equation: **Weir overflow rate** = $\dfrac{\text{Flow, gpd}}{\text{Weir length, ft}}$

First, change 0.95 mgd to gpd.

(0.95 mgd)(1,000,000/1 mil) = 950,000 gpd

Weir overflow rate = $\dfrac{\text{950,000 gpd}}{\text{175 ft}}$ = 5,428.6 gpd/ft, round to **5,400 gpd/ft**

93. **A rectangular clarifier has a weir length of 199 ft. What is the weir overflow rate in gpd/ft, if the flow is 1.6 mgd?**

Equation: **Weir overflow rate** $= \dfrac{\text{Flow, gpd}}{\text{Weir length, ft}}$

First, change 1.6 mgd to gpd.

(1.6 mgd)(1,000,000/1 mil) = 1,600,000 gpd

Weir overflow rate $= \dfrac{1{,}600{,}000 \text{ gpd}}{199 \text{ ft}} = 8{,}040.2$ gpd/ft, round to **8,040 gpd/ft**

94. **A circular clarifier has a diameter of 100.0 ft. If the entire circumference acts as a weir and the flow is 2.88 mgd, what is the weir overflow rate in gpd/ft, if the flow is 2.09 mgd?**

First, change 2.09 mgd to gpd.

(2.09 mgd)(1,000,000/1 mil) = 2,090,000 gpd

Next, determine the weir length. In this case it is the circumference.

Equation: **Circumference** $= \pi$**(Diameter)**

Substitution: Circumference, weir length = 3.14(100.0 ft) = 314 ft

Equation: **Weir overflow rate** $= \dfrac{\text{Flow, gpd}}{\text{Weir length, ft}}$

Weir overflow rate $= \dfrac{2{,}090{,}000 \text{ gpd}}{314 \text{ ft}} =$ **6,656 gpd/ft**

SURFACE OVERFLOW RATE CALCULATIONS

Surface overflow rate determinations are used to determine the loading on clarifiers. The flow amount used in these calculations only counts plant flow, not recirculation.

95. What is the surface overflow rate in gpd/ft², if the basin is 112 ft long, 42 ft wide, and flow into the basin is 1,408,000 gpd?

First, determine the area of the basin.

Area = (Length)(Width)

Area = (112 ft)(42 ft) = 4,704 ft²

Equation: **Surface overflow rate** $= \dfrac{\text{Flow, gpd}}{\text{Area, ft}^2}$

Surface overflow rate $= \dfrac{1,408,000 \text{ gpd}}{4,704 \text{ ft}^2} = 299.3$ gpd/ft², round to **300 gpd/ft²**

96. Given the following data, calculate the surface overflow rate for a clarifier in gpd/ft²:

Clarifier diameter = 74 ft
Primary effluent flow = 1,345,000 gallons per day (gpd)

First, determine the area of the clarifier.

Area $= \pi \mathbf{r}^2$ where r = Diameter/2 = 74 ft/2 = 37 ft

Area = (3.14)(37 ft)² = 4,298.66 ft²

Next, calculate the surface overflow rate.

Equation: **Surface overflow rate** $= \dfrac{\text{Flow, gpd}}{\text{Area, ft}^2}$

Surface overflow rate $= \dfrac{1,345,000 \text{ gpd}}{4,298.66 \text{ ft}^2} = 312.89$ gpd/ft², round to **310 gpd/ft²**

97. **If a sedimentation basin measures 49 ft by 152 ft and receives a flow of 9.048 mgd, what is the surface overflow rate in gallons per day per ft²(gpd/ft²)?**

Equation: **Surface overflow rate** $= \dfrac{\text{Flow, gpd}}{\text{Area, ft}^2}$

First, convert mgd to gpd/ft².

Number of gpd/ft² = (9.048 mgd)(1,000,000 gallons/mil) = 9,048,000 gallons/day

Surface overflow rate $= \dfrac{9,048,000 \text{ gallons/day}}{(49 \text{ ft})(152 \text{ ft})} = 1,214.82$ gpd/ft², round to **1,200 gpd/ft²**

FLOW AND VELOCITY PROBLEMS

Operators need to know the flow and velocity of the wastewater throughout the different plant processes, for example to feed proper dosages of chemicals to treat wastewaters, to know how many clarifiers or ponds to use or how much supernatant to recirculate, and for settling purposes, among other uses.

98. **What is the velocity in feet per second (ft/s) for water flowing through a channel that is 7.2 ft wide, 3.3 ft deep, and the flow is 32.5 ft³/s?**

Equation: **Q (Flow) = (Area)(Velocity)**

Substitute parameters.

32.5 ft³/s = (7.2 ft)(3.3 ft)(Velocity), and solve for velocity by rearranging the equation.

Velocity $= \dfrac{32.5 \text{ ft}^3/\text{s}}{(7.2 \text{ ft})(3.3 \text{ ft})} = 1.368$ ft/s, round to **1.4 ft/s**

99. **What is the flow in ft³/s for a pipe that is 8.0 inches in diameter and the velocity is 1.68 ft/s?** *Note*: **the pipe is flowing full.**

Write the equation: **Q (Flow) = (Area)(Velocity)**

First, convert the diameter to feet.

Know: 12 inches per foot

Diameter, ft = 8.0 in./12 in./ft = 0.667 ft

Next, determine the number of feet for the radius by dividing the diameter by 2.

Radius = 0.667 ft/2 = 0.333 ft

Next, determine the cross-sectional area of the 8-inch diameter pipe.

Area = πr^2, where r is the radius and π equals 3.14

Area = $(3.14)(0.333 \text{ ft})^2$ = 0.348 ft²

Lastly, determine the flow by substituting known values.

Q (flow) = (0.348 ft²)(1.68 ft/s) = **0.58 ft³/s**

100. **What is the velocity of flow in feet per second (ft/s) for a 6.0-inch diameter pipe, if it delivers 204 gpm and is flowing full?**

First, convert the diameter from inches to feet.

Number of ft = (6.0 in.)(1 ft/12 in.) = 0.50 ft

Next, convert the number of gpm to ft³/s.

Number of ft³/s = $\dfrac{204 \text{ gpm}}{(7.48 \text{ gal/ft}^3)(60 \text{ sec}/\text{min})}$ = 0.454 ft³/s

Equation: **Flow, ft³/s = (Area, ft²)(Velocity, ft/s)** where the Area = $(0.785)(D)^2$ or πr^2

0. 454 ft³/s = (0.785)(0.50)(0.50)(Flow, ft/s)

Rearrange and solve for the flow in ft/s.

Flow, ft/s = $\dfrac{0.454 \text{ ft}^3/\text{s}}{(0.785)(0.50 \text{ ft})(0.50 \text{ ft})}$ = 2.31, round to **2.3 ft/s**

101. **Water flowing through a channel is 8.4 ft wide and 3.6 ft deep. If the flow is 20.3 ft³/s, what is the velocity of the water?**

Equation: **Q (Flow) = (Area)(Velocity)**

$20.3 \text{ ft}^3/\text{s} = (8.4 \text{ ft})(3.6 \text{ ft})(\text{Velocity, ft/s})$

Now, solve for velocity.

$$\text{Velocity, ft/s} = \frac{20.3 \text{ ft}^3}{(8.4 \text{ ft})(3.6 \text{ ft})} = \textbf{0.67 ft/s}$$

PUMP DISCHARGE PROBLEMS

Operators need to understand pump discharge calculations, for example to plan treatment processes and time, to determine how long a pump will take to discharge a certain amount of wastewater or chemical to treat the wastewater, and maybe to change the size of a pump to fit the need better.

102. **How long will it take in hours for a pump to discharge 265,000 gallons, if it is pumping at the rate equal to 450 gal/min?**

First, determine the number of minutes it will take.

$$\frac{265,000 \text{ gal}}{450 \text{ gal/min}} = 588.88 \text{ min}$$

Then, convert to hours by dividing by 60 minutes/hour.

$$\frac{588.88 \text{ min}}{60 \text{ min/hr}} = \textbf{9.8 hours}$$

103. A pump's output is averaging 450 gallons/min. How many gallons will it pump in one day?

(450 gal/min)(1,440 min/day) = 648,000 gal/day, round to **650,000 gal/day**

104. If a pump discharges 40,900 gallons in 3 hours and 5 minutes, how many gallons/ minute is the pump discharging?

First, determine the number of minutes the pump was working.

3 hr(60 min/hr) + 5 min = 180 min + 5 min = 185 min

Then, determine the number of gallons/min by dividing the number of gallons pumped by the total time the pump worked.

$$\frac{40,900 \text{ gal}}{185 \text{ min}} = \textbf{221 gal/min}$$

105. Determine a pump's total output in mgd, if it is pumping 375 gal/min.

(375 gal/min)(1,440 min/day)(1 mil gal/1,000,000) = **0.540 mgd**

106. How many gallons will a pump discharge, if it pumps an average of 150 gallons/ min for 3 hours and 12 minutes?

First, calculate the number of minutes for the time interval given.

3 hr(60 min/hr) + 12 min = 180 min + 12 min = 192 min

Then, calculate the number of gallons by multiplying the length of time the pump worked in minutes times the number of gallons/min.

(192 min)(150 gal/min) = 28,800 gal, round to **29,000 gal**

107. **A meter that reads in hundreds of gallons on the discharge side of a pump is read each morning. If it read 713,436 at 8:15 AM on Tuesday morning and the next day read 714,852 at 7:30 AM, what is the average gpm that is discharged from this pump for that time period?**

First, determine the number of gallons pumped for this time period.

Number of gal = (714,852 − 713,436)(100 gal) = 141,600 gal

Next, determine the number of minutes between reads.

Since the second read is 45 minutes before the first read it follows:

Number of minutes = 1,440 min − 45 min = 1,395 min

Since there are 1,440 min/day.

Now, calculate the pumping rate in gpm.

$$\text{Pumping rate, gpm} = \frac{141,600 \text{ gal}}{1,395 \text{ min}} = 101.5 \text{ gpm, round to } \textbf{102 gal}$$

SOLIDS AND HYDRAULIC LOADING RATE PROBLEMS

Solids and hydraulic loading rate calculations are used to determine the solids or hydraulic loading on clarifiers, trickling filters, and other processes. These calculations are important to know so operators, for example, can determine when to discharge sludge from a clarifier or to know the contact time between organisms in a trickling filter and the food entering that trickling filter.

108. **Given the following parameters, calculate the solids loading rate on a secondary clarifier:**

Clarifier diameter = 90.3 ft
Activated sludge flow = 2.88 mgd
Mixed liquor suspended solids (MLSS) = 3,850 mg/L

First, determine the area of the clarifier.

Area $= \pi r^2$ where r $=$ Diameter/2 $=$ 90.3 ft/2 $=$ 45.15 ft

Area $=$ (3.14)(45.15 ft)^2 $=$ 6,401 ft^2

Next, calculate the solids loading rate.

Equation: **Solids loading rate** $= \dfrac{(\text{MLSS, mg/L})\,(\text{mgd})\,(8.34\ \text{lb/gal})}{\text{Area, ft}^2}$

Solids loading rate, lb of solids/d/ft^2 $= \dfrac{(3{,}850\ \text{mg/L})\,(2.88\ \text{mgd})\,(8.34\ \text{lb/gal})}{6{,}401\ \text{ft}^2}$

Solids loading rate, lb of solids/d/ft^2 $=$ **14.4 lb of solids/d/ft^2**

109. What is the hydraulic loading rate on a trickling filter in gallons per day per square foot (gpd/ft^2) given the following data?

Flow = 3.95 mgd
Diameter of tricking filter = 100.2 ft
Clarifier recirculation rate = 0.38 mgd

First, determine the total flow in gallons per day (gpd) through the trickling filter.

Total flow, gal $=$ (3.95 mgd $+$ 0.38 mgd)(1,000,000 gal/mil) $=$ 4,330,000 gpd

Next, determine the surface area in ft^2 for the trickling filter.

Area $= \pi r^2$ where r $=$ Diameter/2 $=$ 100.2 ft/2 $=$ 50.1 ft

Trickling filter surface area, ft^2 $=$ (3.14)(50.1 ft)(50.1 ft) $=$ 7,881.43 ft^2

Lastly, calculate the hydraulic loading rate.

Hydraulic loading rate $= \dfrac{\text{Total flow, gpd}}{\text{Surface area, ft}^2}$

Hydraulic loading rate $= \dfrac{4{,}330{,}000,\ \text{gpd}}{7{,}881.43\ \text{ft}^2} =$ 549 gpd/ft^2, round to **550 gpd/ft^2**

110. **A secondary clarifier has a radius of 42.5 ft. If the MLSS is 3,125 mg/L, the secondary effluent flow is 2.3 mgd, and the return of activated sludge is 0.48 mgd, what is the solids loading rate?**

First, determine the total flow.

Total flow = Primary flow + Return of activated sludge

Total flow = 2.3 mgd + 0.48 mgd = 2.78 mgd

Next, calculate the area of the clarifier.

Area = πr^2

Area = (3.14)(42.5 ft)(42.5 ft) = 5,671.625 ft²

Next, calculate the solids loading rate.

Equation: **Solids loading rate** $= \dfrac{(\text{MLSS, mg/L})(\text{mgd})(8.34 \text{ lb/gal})}{\text{Area, ft}^2}$

Solids loading rate $= \dfrac{(3,125 \text{ mg/L})(2.78 \text{ mgd})(8.34 \text{ lb/gal})}{5,671.625 \text{ ft}^2}$

Solids loading rate = 12.77 lb of solids/d/ft², round to **13 lb of solids/d/ft²**

SLUDGE PUMPING PROBLEMS

Sludge pumping calculations are important for operators to determine so they know how much sludge and solids are being loaded into a digester to prevent underloading or overloading of the digester. Also, operators need to know how much sludge is being pumped to other sludge processing applications such as sludge thickening, filter presses, or land application.

111. **A sludge pumps operates for exactly 10 minutes every hour and pumps an average of 26.2 gpm of sludge from the settling tank to the digester. If the percent solids in the sludge averages 4.1%, determine the amount of sludge pumped in pounds per day.**

Solids, lb/day = (10 min/hr)(24 hr/day)(26.2 gpm)(8.34 lb/gal)(4.1%/100%)

Solids, lb/day = 2,150.12 lb/day, round to **2,200 lb/day**

112. **A sludge pump operates for 12 minutes every hour and pumps an average of 30.7 gpm of sludge from the settling tank to the digester. If the percent solids in the sludge averages 4.9% and the volatiles are 64%, determine the amount of volatiles pumped in pounds per day.**

Volatiles, lb/day = (12 min/hr)(24 hr/day)(30.7 gpm)(8.34 lb/gal)(4.9 %/100 %)(64%/100%)

Volatiles, lb/day = 2,312.45 lb/day, round to **2,300 lb/day**

BIOSOLIDS PUMPING AND PRODUCTION PROBLEMS

Biosolids pumping calculations provide operators accurate process control data for the sedimentation process. Biosolids are mostly composed of water with the biosolids ranging from only 3 to 7% by volume.

113. **Assuming sludge is 8.34 lb/gal, what is the estimated biosolids pumping rate in gpm for the following wastewater system?**

Plant flow = 0.925 mgd
Removed biosolids = 1.14%
Influent total suspended solids (TSS) = 341 mg/L
Effluent TSS = 125 mg/L

Equation: **Estimated pumping rate =**

$$\frac{(\text{Influent TSS, mg/L} - \text{Effluent TSS, mg/L})(\text{Flow, mgd})(8.34\,\text{lb/gal})}{(\text{Percent solids in sludge})(\text{Sludge, lb/gal})(1,440\,\text{min/day})}$$

Substitute values and solve.

$$\text{Estimated pumping rate} = \frac{(341\,\text{TSS mg/L} - 125\,\text{TSS, mg/L})(0.925\,\text{mgd})(8.34\,\text{lb/gal})}{(1.14\%/100\%)(8.34\,\text{lb/gal})(1,440\,\text{min/day})}$$

$$\text{Estimated pumping rate} = \frac{(216\,\text{TSS mg/L})(0.925\,\text{mgd})(8.34\,\text{lb/gal})}{(1.14\%/100\%)(8.34\,\text{lb/gal})(1,440\,\text{min/day})}$$

Estimated pumping rate = **12.2 gpm**

114. What is the biosolids production in lb/mil gal, if a wastewater treatment plant produces 74,300 gallons of biosolids in a 28-day month and the plant treated 1.15 mgd on average?

Equation: **Biosolids, lb/mil gal** $= \dfrac{(\text{Biosolids, gal/day})(8.34 \text{ lb/gal})}{(\text{Flow, mgd})(\text{Number of days})}$

Biosolids, lb/mil gal $= \dfrac{(74,300 \text{ gal/day})(8.34 \text{ lb/gal})}{(1.15 \text{ mgd})(28 \text{ days})}$

Biosolids, lb/mil gal $=$ 19,244 lb/mil gal, round to **19,200 lb/mil gal**

115. If the plant flow averages 1.38 mgd and production of biosolids averages 13,950 gal/day, what is the biosolids production in wet tons per year? A material in its natural state, wet state, is measured in wet tons.

Equation: **Biosolids, wet tons/yr** $= \dfrac{(\text{Biosolids, lb/mil gal})(\text{mgd})(365 \text{ days/yr})}{2,000 \text{ lb/ton}}$

Substitute values and solve.

Biosolids, wet tons/yr $= \dfrac{(13,950 \text{ lb/mil gal})(1.38 \text{ mgd})(365 \text{ days/yr})}{2,000 \text{ lb/ton}}$

Biosolids, wet tons/yr $=$ 3,513 wet tons/yr, round to **3,510 wet tons/yr**

116. What is the estimated biosolids pumping rate for the following system?

Plant flow = 3.16 mgd
Removed biosolids = 1.26%
Influent total suspended solids (TSS) = 294 mg/L
Effluent TSS = 131 mg/L
Biosolids weigh = 8.42 lb/gal

Equation: **Estimated pumping rate** $=$

$\dfrac{(\text{Influent TSS, mg/L} - \text{Effluent TSS, mg/L})(\text{Flow, mgd})(8.34 \text{ lb/gal})}{(\text{Percent solids in sludge})(\text{Sludge, lb/gal})(1,440 \text{ min/day})}$

Substitute values and solve.

$$\text{Estimated pumping rate} = \frac{(294 \text{ TSS mg/L} - 131 \text{ TSS, mg/L})(3.16 \text{ mgd})(8.34 \text{ lb/gal})}{(1.26\%/100\%)(8.42 \text{ lb/gal})(1,440 \text{ min/day})}$$

$$\text{Estimated pumping rate} = \frac{(163 \text{ TSS mg/L})(3.16 \text{ mgd})(8.34 \text{ lb/gal})}{(1.26\%/100\%)(8.42 \text{ lb/gal})(1,440 \text{ min/day})}$$

Estimated pumping rate = **28.1 gpm**

WASTE ACTIVATED SLUDGE PUMPING RATE CALCULATIONS

These calculations are used as a planning tool by the operator. The waste activated sludge (WAS) suspended solids (SS) are pumped out of the secondary clarifier and wasted or returned to the aeration tank. It is better to pump continuously rather than intermittently and not to change the amount by more than 15% from one day to the next.

117. Determine the waste activated sludge (WAS) pumping rate in gpm, if 3,200 lb/ day are to be wasted and the WAS suspended solids concentrations, 3,980 mg/L.

Use the "pounds" equation to solve this problem.

Equation: **Number of lb/day WAS = (WAS, mg/L)(Number of mgd)(8.34 lb/gal)**

Rearrange the equation to solve for mg/L.

$$\textbf{Number of mgd} = \frac{\textbf{Number of lb/day WAS}}{(\textbf{Number of mg/L WAS})(8.34 \text{ lb/gal})}$$

Substitute values and solve.

$$\text{Number of mgd} = \frac{3,200 \text{ lb/day}}{(3,980 \text{ mg/L WAS})(8.34 \text{ lb/gal})} = \textbf{0.0964 mgd}$$

Lastly, convert mgd to gpm.

$$\text{WAS pumping rate, gpm} = \frac{(0.0964 \text{ mgd})(1,000,000/\text{mil})}{1,440 \text{ min/day}} = \textbf{67 gpm}$$

118. Determine the waste activated sludge (WAS) pumping rate in gpm given the following data:

Amount of WAS to be wasted = **4,800 lb/day**
WAS suspended solids concentrations = **4,125 mg/L**

Use the "pounds" equation to solve this problem.

Equation: **Number of lb/day WAS = (WAS, mg/L)(Number of mgd)(8.34 lb/gal)**

Rearrange the equation to solve for mg/L.

$$\textbf{Number of mgd} = \frac{\text{Number of lb/day WAS}}{(\text{Number of mg/L WAS})(8.34\ \text{lb/gal})}$$

Substitute values and solve.

$$\text{Number of mgd} = \frac{4,800\ \text{lb/day}}{(4,125\ \text{mg/L WAS})(8.34\ \text{lb/gal})} = 0.1395\ \text{mgd}$$

Lastly, convert mgd to gpm.

$$\text{WAS pumping rate, gpm} = \frac{(0.1395\ \text{mgd})(1,000,000/\text{mil})}{1,440\ \text{min/day}} = \textbf{97 gpm}$$

TOTAL DYNAMIC HEAD CALCULATIONS

These calculations tell the operator how hard a pump has to work against the static head (the height above the pump that the liquid is discharged to), as well as head losses caused from friction from the water and pipe contact, bends in the pipe, and appurtenances (valves or orifices) along the pipeline. For simplicity this book will refer to total dynamic head (TDH in some literature) as total head.

119. Find the total head in ft for a pump with a total static head of 145 ft and a head loss of 5.8 ft.

Equation: **Total head, ft = Total static head, ft + Head losses, ft**

Total head, ft = 145 ft + 5.8 ft = 150.8 ft, round to **150 ft**

120. **Water is being pumped from a water source with an elevation of 378 ft to an elevation of 645 ft. What is the total head, if friction and minor head losses are 17.5 ft?**

Equation: **Total head, ft = Total static head, ft (difference in elevation) + Head losses, ft**

Total head, ft = (645 ft − 378 ft) + 17.5 ft = 267 + 17.5 ft = 284.5 ft, round to **284 ft**

121. **Water is being pumped from a water source with an elevation of 845 ft to an elevation of 1,033 ft. What is the total head, if friction and minor head losses are 13.5 ft?**

Equation: **Total head, ft = Total static head, ft (difference in elevation) + Head losses, ft**

Total head, ft = (1,033 ft − 845 ft) + 13.5 ft = 188 + 13.5 ft = 201.5 ft, round to **202 ft**

122. **A lift pump at an elevation of 22.4 ft is pumping to a wastewater collection pond that is at an elevation of 41.7 ft. If minor head losses are 6.2 ft, what is the total dynamic head (TDH)?**

Equation: **Total head, ft = Total static head, ft (difference in elevation) + Head losses, ft**

Total head, ft = (41.7 ft − 22.4 ft) + 6.2 ft = 19.3 + 6.2 ft = **25.5 ft**

PUMPING HORSEPOWER AND EFFICIENCY CALCULATIONS

These types of calculations can be used for determining pump size, efficiency, and costing.

123. **Find the water horsepower (whp), if the brake horsepower (bhp) is 70.7 and the pump efficiency is 66%.**

Equation: **Water horsepower = (Brake horsepower)(Pump efficiency)**

Water hp = (70.7)(66%/100% Pump efficiency) = 46.662 whp, round to **47 whp**

124. What is the brake horsepower (bhp), if the water horsepower (whp) is 88 and the pump efficiency is 71%?

Equation: **Brake hp = whp/Pump efficiency**

$$\text{Brake hp} = \frac{88 \text{ whp}}{71\%/100\% \text{ Pump efficiency}} = 123.9 \text{ bhp, round to } \textbf{120 bhp}$$

125. What is the motor horsepower (mhp), if 275 horsepower (hp) is required to run a pump with a motor efficiency (ME) of 88% and a pump efficiency (PE) of 73%? Note: The 275 hp in this problem is called the water horsepower (whp). The whp is the actual energy (horsepower) available to pump water.

Equation: **Motor horsepower** $= \dfrac{\text{whp}}{(\text{ME})(\text{PE})}$

$$\text{mhp} = \frac{275 \text{ whp}}{(88\%/100\% \text{ ME})(73\%/100\% \text{ PE})}$$

$$\text{mhp} = \frac{275 \text{ whp}}{(0.88 \text{ ME})(0.73 \text{ PE})}$$

mhp = 428.08 mhp, round to **430 mhp**

126. What is the motor horsepower, if the brake horsepower (bhp) is 81 and the motor efficiency (ME) is 86%?

Equation: **Motor horsepower = Brake horsepower/Motor efficiency**

$$\text{Motor horsepower (mhp)} = \frac{81 \text{ bhp}}{86\%/100\% \text{ ME}} = \textbf{94 mhp}$$

127. What is the cost to run a pump in dollars and cents per day, if the horsepower is 149 and the cost per kW-hr is $0.083?

Equation: **Cost, $/day = (Motor hp)(24 hr/day)(0.746 kW/hp)(Cost/kW-hr)**

Cost, $/day = (149 hp)(24 hr/day)(0.746 kW/hp)($0.083/kW-hr) = **$221.42/day**

128. What is the brake horsepower (bhp), if 180 horsepower (hp) is supplied to a motor with 91% efficiency?

Equation: **Brake horsepower = (Horsepower)(Motor efficiency)**

Brake hp = (180 hp)(91%/100% Motor efficiency) = 163.8 bhp, round to **160 bhp**

129. Find the water horsepower, if the brake horsepower (bhp) is 96 and the pump efficiency is 78%.

Equation: **Water horsepower = (Brake horsepower)(Pump efficiency)**

Water horsepower = (96 bhp)(78%/100% Pump efficiency) = 74.88 whp, round to **75 whp**

130. What is the motor horsepower (mhp), if 240 horsepower (hp = water horsepower or whp) is required to run a pump with a motor efficiency (ME) of 92% and a pump efficiency (PE) of 82%?

Equation: **Motor horsepower $= \dfrac{\text{whp}}{\text{(ME)(PE)}}$**

$$\text{mhp} = \frac{240 \text{ whp}}{(92\%/100\% \text{ ME})(82\%/100\% \text{ PE})}$$

$$\text{mhp} = \frac{240 \text{ whp}}{(0.92 \text{ ME})(0.82 \text{ PE})}$$

mhp = 318.13 mhp, round to **320 mhp**

131. Find the water horsepower, if the brake horsepower is 67 and the pump efficiency is 68%.

Equation: **Water horsepower = (Brake horsepower)(Pump efficiency)**

Water horsepower = (67)(68%/100% Pump efficiency) = 45.56 whp, round to **46 whp**

132. **What must have been the pumping efficiency (PE) for a pump given the following data?**

Water horsepower (whp) = 259
Motor horsepower (mhp) = 365
Motor efficiency (ME) = 89.5%

Equation: **Motor horsepower** $= \dfrac{\text{whp}}{(\text{ME})(\text{PE})}$

Rearrange the equation to solve for pump efficiency. *Note*: Remember the motor and pump efficiencies are divided by 100% to convert them to the proper units (see above problems), thus, the reason for the 100% in the numerator and the denominator.

Pump efficiency $= \dfrac{(\text{whp})(100\%)}{(\text{mhp})(\text{ME}/100\%)}$

Substitute values and solve.

Pump efficiency $= \dfrac{(259)(100\%)}{(365)(89.5\%/100\%)} = 79.28\%$, round to **79.3% Pump efficiency**

133. **What is the pump efficiency for a pump that has a brake horsepower of 180 and a water horsepower of 140?**

Equation: **Water horsepower = (Brake horsepower)(Pump efficiency)**

Rearrange the equation to solve for pump efficiency.

Pump efficiency = (Water horsepower/brake horsepower)(100%)

Substitute values and solve.

Pump efficiency $= \dfrac{(140)(100\%)}{180} = 77.77\%$, round to **78% Pump efficiency**

CALCULATIONS OF CHLORINE DOSE, DEMAND, AND RESIDUAL

Usually used by operators to determine chlorine demand because they can calculate the dose and measure the residual.

134. If the chlorine dose is 12.2 mg/L and the chlorine residual is 0.5 mg/L, what is the chlorine demand?

Equation: **Chlorine dose = Chlorine demand + Chlorine residual**

Rearrange and solve for chlorine demand.

Chlorine demand = Chlorine dose − Chlorine residual

Chlorine demand = 12.2 mg/L − 0.5 mg/L = **11.7 mg/L**

135. If the chlorine dose is 8.05 mg/L and the chlorine demand is 7.43 mg/L, what is the chlorine residual?

Equation: **Chlorine dose = Chlorine demand + Chlorine residual**

Rearrange and solve for chlorine residual.

Chlorine residual = Chlorine dose − Chlorine demand

Chlorine residual = 8.05 mg/L − 7.43 mg/L = **0.62 mg/L**

136. **If the chlorine residual is 0.55 mg/L and the chlorine demand is 7.02 mg/L, what is the chlorine dose?**

Equation: **Chlorine dose = Chlorine demand + Chlorine residual**

Chlorine dose = 7.02 mg/L + 0.55 mg/L = **7.57 mg/L**

DOSAGE PROBLEMS

These calculations are used mainly for process control, which requires accurate determination before the chemical is actually applied to a particular process. By keeping accurate records of dosages and thus usage, operators can also plan ordering or costing.

137. **How many gpd of a 9.5% sodium hypochlorite solution are needed to disinfect a flow of 635,000 gallons, if the dosage required is 9.75 mg/L? Assume the solution weighs 8.34 lb/gal.**

First, convert gpd to mgd.

$$\text{Number of mgd} = \frac{635,000 \text{ gpd}}{1,000,000/\text{mil}} = 0.635 \text{ mgd}$$

Next, using the "pounds equation," calculate the lb/day of chlorine needed.

Equation: Chlorine, lb/day = (Dosage, mg/L)(mgd)(8.34 lb/gal)

Chlorine, lb/day = (9.75 mg/L)(0.635 mgd)(8.34 lb/gal) = 51.635 lb/day

Since the solution is not 100%, divide the percent hypochlorite into the lb/day of chlorine needed.

$$\text{Hypochlorite, lb/day} = \frac{51.635 \text{ lb/day}}{9.5\%/100\%} = 543.53 \text{ lb/day hypochlorite}$$

Lastly, determine the gpd of hypochlorite solution needed.

$$\text{Hypochlorite, gpd} = \frac{543.53 \text{ lb/day}}{8.34 \text{ lb/gal}} = 65.17 \text{ gpd, round to } \textbf{65 gpd Sodium hypochlorite}$$

138. **How many lb/day of chlorine gas is required to treat 3.44 mgd, if the desired dosage is 7.0 mg/L?**

Equation: **Number of lb/day = (Cl$_2$, mg/L)(Number of mgd)(8.34 lb/gal)**

Substitute values and solve.

Number of lb/day Cl$_2$ = (7.0 mg/L)(3.44 mgd)(8.34 lb/gal)

Number of lb/day Cl$_2$ = 200.82 lb/day, round to **200 lb/day Cl$_2$**

139. **How much sulfur dioxide in lb/day needs to be applied to dechlorinate a wastewater, if the flow is 2.11 mgd, the chlorine residual is 1.15 mg/L, and the sulfur dioxide must be 3.0 mg/L higher than the chlorine residual?**

First, determine how many mg/L of sulfur dioxide must be applied. This is the chlorine residual plus the amount that is required higher than the chlorine residual. This additional amount of sulfur dioxide above the chlorine residual is applied as a safety factor and is typically started at 3 mg/L.

SO$_2$, mg/L = 1.15 mg/L + 3.0 mg/L = 4.15 mg/L SO$_2$

Next, determine the number of lb/day of SO$_2$ needed.

Equation: **Number of lb/day SO$_2$ = (SO$_2$, mg/L)(Number of mgd)(8.34 lb/gal)**

Substitute values and solve.

Number of lb/day SO$_2$ = (4.15 mg/L)(2.11 mgd)(8.34 lb/gal)

Number of lb/day SO$_2$ = 73.03 lb/day, round to **73 lb/day SO$_2$**

140. **Determine the number of lb/day of SO_2 used to treat a wastewater plant's effluent given the flowing data:**

Average flow per day = 950,000 gallons
Average chlorine residual = 1.42 mg/L
SO_2 dosage 3.0 mg/L higher than chlorine residual

First, convert gallons per day to mgd.

Number of mgd = (950,000 gpd) / (1,000,000/mil) = 0.95 mgd

Next, determine the mg/L of SO_2 that must be applied.

Number of mg/L SO_2 = 1.42 mg/L + 3.0 mg/L = 4.42 mg/L SO_2

Lastly, determine the lb/day SO_2.

Equation: **Number of lb/day SO_2 = (SO_2, mg/L)(Number of mgd)(8.34 lb/gal)**

Substitute values and solve.

Number of lb/day SO_2 = (4.42 mg/L)(0.95 mgd)(8.34 lb/gal) = **35 lb/day SO_2**

141. **What is the number of lb/day of alum used by a wastewater plant given the following data?**

Plant's treatment flow = 1,225 gpm
Alum dose = 13.85 mg/L
Alum = 48.5% aluminum sulfate

First, convert gpm to mgd.

$$\text{Number of mgd} = \frac{(1,225 \text{ gpm})\,(1,440 \text{ min/day})}{1,000,000/\text{mil}} = 1.764 \text{ mgd}$$

Next, calculate the number of lb/day of alum required.

Equation: $\textbf{lb/day} = \dfrac{(\text{Alum, mg/L})\,(\text{mgd})\,(8.34\ \text{lb/gal})}{\text{Percent purity}/100\%}$

Alum, lb/day $= \dfrac{(13.85\ \text{mg/L})\,(1.764\ \text{mgd})\,(8.34\ \text{lb/gal})}{48.5\%/100\%} = \textbf{420 lb/day of Alum}$

142. **A wastewater plant is treating 1.33 mgd at a chlorine dosage of 9.2 mg/L. If the sodium hypochlorite being used is 14.5% available chlorine, what is the chlorine usage in lb/day?**

Equation: $\textbf{Chlorine, lb/day} = \dfrac{(\text{Dosage, mg/L})\,(\text{mgd})\,(8.34\ \text{lb/gal})}{\text{Percent available chlorine}/100\%}$

Substitute values and solve.

Chlorine, lb/day $= \dfrac{(9.2\ \text{mg/L})\,(1.33\ \text{mgd})\,(8.34\ \text{lb/gal})}{14.5\%/100\%\ \text{Available chlorine}}$

Chlorine, lb/day = 703.78 lb/day, round to **700 lb/day Sodium hypochlorite**

CHEMICAL FEED SOLUTION SETTINGS

Again, these calculations are used mainly for process control, which requires accurate determination before the chemical is actually applied to a particular process. And again, by keeping accurate records of dosages and thus usage, operators can also plan ordering or costing.

143. **If a polymer pump is delivering 12.3 gpd, what is the feed rate in mL/min?**

Equation: $\textbf{Feed rate, mL/min} = \dfrac{(\text{gpd})\,(3{,}785\ \text{mL/gal})}{1{,}440\ \text{min/day}}$

Feed rate, mL/min $= \dfrac{(12.3\ \text{gpd})\,(3{,}785\ \text{mL/gal})}{1{,}440\ \text{min/day}} = \textbf{32.3 mL/min}$

144. **What should the chemical feeder be set on in mL/min, if the desired polymer dosage is 35.9 gpd?**

Equation: **Number of mL/min** $= \dfrac{(\text{Number of gallons used})\,(3{,}785\ \text{mL/gal})}{1{,}440\ \text{min/day}}$

Substitute values and solve.

Polymer, mL/min $= \dfrac{(35.9\ \text{gal})\,(3{,}785\ \text{mL/gal})}{1{,}440\ \text{min/day}} = 94.36\ \text{mL/min, round to}\ \textbf{94 mL/min}$

145. **What should the chemical feeder be set on in mL/min, if the desired polymer dosage is 27.5 gpd?**

Equation: **Number of mL/min** $= \dfrac{(\text{Number of gallons used})\,(3{,}785\ \text{mL/gal})}{1{,}440\ \text{min/day}}$

Substitute values and solve.

Polymer, mL/min $= \dfrac{(27.5\ \text{gal})\,(3{,}785\ \text{mL/gal})}{1{,}440\ \text{min/day}} = 72.28\ \text{mL/min, round to}\ \textbf{72.3 mL/min}$

146. **A wastewater treatment plant uses alum that is 5.37 lb of dry alum per gallon of solution. If it has been determined that 13.5 mg/L of alum is optimum, what should the chemical feed pump be set on in gpd, if the plant is treating 1.25 mgd?**

First, calculate the number of lb/day of alum required.

Equation: **lb/day = (Alum dose, mg/L)(mgd)(8.34 lb/gal)**

Alum, lb/day = (13.5 mg/L)(1.25 mgd)(8.34 lb/gal) = 140.74 lb/day of dry alum

Now, calculate the amount of liquid alum by dividing the amount of dry alum by 5.37 lb/gal.

Alum, gpd $= \dfrac{140.74\ \text{lb/day}}{5.37\ \text{lb/gal}} = \textbf{26.2 gpd, of Liquid alum solution}$

DRY CHEMICAL FEED SETTINGS

As with liquid dosing, accuracy in dosing dry chemicals is important too. The more accurate the dosage calculation is, the more probability there will be for an operator to control a treatment process and the better the records for future referral.

147. **Determine the feed rate of dry alum in lb/day, if the drawdown in exactly 5 minutes was 126.4 grams (g) and the flow is 1.77 mgd.**

First, determine the number of grams used per minute.

Alum, g = 126.4 g/5 min = 25.28 g/min

Know: 454 grams = 1 pound

Equation: $\textbf{Alum, lb/day} = \dfrac{(\text{Number of g/min})(1,440 \text{ min/day})}{454 \text{ g/lb}}$

Substitute values and solve.

$\text{Alum, lb/day} = \dfrac{(25.28 \text{ g/min})(1,440 \text{ min/day})}{454 \text{ g/lb}} = 80.18 \text{ lb/day, round to } \textbf{80.2 lb/day Alum}$

148. **What must have been the setting of a gravimetric dry chemical feeder in g/min, if the number of lb/day was 66.2?**

Equation: $\textbf{Chemical, lb/day} = \dfrac{(\text{Number of g/min})(1,440 \text{ min/day})}{454 \text{ g/lb}}$

Rearrange to solve for the feeder setting in g/min.

$\text{Number of g/min} = \dfrac{(\text{Chemical, lb/day})(454 \text{ g/lb})}{1,440 \text{ min/day}}$

Substitute values and solve.

$\text{Number of g/min} = \dfrac{(66.2 \text{ lb/day})(454 \text{ g/lb})}{1,440 \text{ min/day}} = \textbf{20.9 g/min}$

149. **What is the feed rate of a dry chemical in lb/day, if a sample collection tray collected 93 g in exactly 30 minutes?**

First, determine the number of g/min.

Number of grams = 93 g/30 min = 3.1 g/min

Equation: **Chemical, lb/day** $= \dfrac{(\text{Number of g/min})\,(1{,}440\ \text{min/day})}{454\ \text{g/lb}}$

Substitute values and solve.

Chemical, lb/day $= \dfrac{(3.1\ \text{g/min})\,(1{,}440\ \text{min/day})}{454\ \text{g/lb}} =$ **9.8 lb/day**

SLUDGE PRODUCTION CALCULATIONS

Sludge production calculations are important for costing and disposal purposes. Plants that use processes like digestion or heat treatment have smaller sludge production because more of the sludge is destroyed compared to plants that use chemical addition to treat wastes.

150. **Given the following data, determine the amount of primary dry solids produced in lb/day:**

Flow = 2.88 mgd
Influent suspended solids removed = 297 mg/L
Primary effluent suspended solids = 126 mg/L

First, determine the number of mg/L of suspended solids (SS) removed.

SS removed, mg/L = 297 mg/L, influent − 126 mg/L, effluent = 171 mg/L SS removed

SS removed, lb/day = (SS removed, mg/L)(Number of mgd)(8.34 lb/gal)

SS removed, lb/day = (171 mg/L SS)(2.88 mgd)(8.34 lb/gal)

SS removed, lb/day = 4,107.28 lb/day, round to **4,110 lb/day**

151. A wastewater plant with an influent flow of 1.79 mgd has primary influent suspended solids of 209 mg/L. If the secondary suspended solids are 101 mg/L, what is the amount of dry solids produced in lb/day?

First, determine the number of mg/L of suspended solids (SS) removed.

SS removed, mg/L = 209 mg/L, influent − 101 mg/L effluent = 108 mg/L of SS removed

SS removed, lb/day = (SS removed, mg/L)(Number of mgd)(8.34 lb/gal)

SS removed, lb/day = (108 mg/L SS)(1.79 mgd)(8.34 lb/gal)

SS removed, lb/day = 1,612.29 lb/day, round to **1,610 lb/day**

152. Given the following data, determine the amount of flow the wastewater plant is treating.

Primary effluent suspended solids = 127 mg/L
Primary effluent suspended solids removed = 2,342 lb/day

Equation: **SS removed, lb/day = (SS removed, mg/L)(Number of mgd)(8.34 lb/gal)**

Rearrange to solve for mg/L of SS removed.

$$\text{Number of mgd} = \frac{\text{SS removed, lb/day}}{(\text{SS removed, mg/L})(8.34 \text{ lb/gal})}$$

$$\text{Number of mgd} = \frac{2{,}342 \text{ lb/day}}{(127 \text{ mg/L SS})(8.34 \text{ lb/gal})} = \textbf{2.21 mgd}$$

SLUDGE-AGE (GOULD) CALCULATIONS

Operators need to understand sludge-age calculations because this will help them maintain an appropriate amount of activated sludge in an aeration tank. The age of the sludge refers to the average solids retention time (usually in days) that the solids remain in the aeration tank. The sludge age is controlled by the sludge wasting rate, which affects the sludge yield in the system. This calculation is similar to detention time. See Figure 8 in Appendix E for one type of wastewater plant using an oxidation ditch.

153. **A wastewater treatment plant adds 408 lb/day of solids to its oxidation ditch (aeration lagoon). If the solids under aeration are 6,750 lb, what is the sludge age in days?**

Equation: $\textbf{Sludge age, days} = \dfrac{\text{Solids under aeration, lb}}{\text{Solids added, lb/day}}$

$\text{Sludge age, days} = \dfrac{6,750 \text{ lb}}{408 \text{ lb/day}} = 16.54 \text{ day, round to } \textbf{16.5 days}$

154. **How much solids must be added per day to a wastewater plant's oxidation ditch, if the sludge age is 20.4 days and the solids under aeration are 2,460 lb?**

Equation: $\textbf{Sludge age, days} = \dfrac{\text{Solids under aeration, lb}}{\text{Solids added, lb/day}}$

Rearrange the equation to solve for solids added in lb/day.

$\text{Solids added, lb/day} = \dfrac{\text{Solids under aeration, lb}}{\text{Sludge age, days}}$

Substitute values and solve.

$\text{Solids added, lb/day} = \dfrac{2,460 \text{ lb}}{20.4 \text{ days}} = 120.59 \text{ lb/day, round to } \textbf{121 lb/day}$

SLUDGE VOLUME INDEX AND SLUDGE VOLUME DENSITY CALCULATIONS

The sludge volume index (SVI) and density calculations inform the operator about the flocculation process, the settling characteristics of the activated sludge, and how these settling characteristics affect return sludge rates and mixed liquor suspended solids. Activated sludge plants that are functioning properly usually have an SVI of around 100 mL/g.

155. A settleability test on MLSS sample in a 1-liter graduated cylinder had a suspended solids result of 248 mL. If the mixed liquor suspended solids (MLSS) was 2,620 mg/L in the aeration tank, what was the sludge volume index (SVI)? What does the result tell you?

First, convert MLSS in mg to grams.

MLSS, g = (2,620 mg)(1 g/1,000 mg) = 2.62 g

Equation: $\mathbf{SVI} = \dfrac{\text{SS, mL}}{\text{MLSS, g/L}}$

Substitute values and solve.

$SVI = \dfrac{248 \text{ mL}}{2.62 \text{ g}} = 94.66$, round to **94.7 SVI**

The result indicates the biosolids are old and effluent turbidity is probably increasing.

156. A 30-minute settleability test on MLSS sample had a settleable solids result of 286 mL in a 1-liter graduated cylinder. If the mixed liquor suspended solids (MLSS) in the aeration tank were 3,025 mg/L, what was the sludge density index (SDI)?

First, convert MLSS in mg to grams.

MLSS, g = (3,025 mg)(1 g/1,000 mg) = 3.025 g, MLSS

Equation: $\mathbf{SDI} = \dfrac{(\text{MLSS, g})(100\%)}{\text{SS, mL}}$

Substitute values and solve.

$SDI = \dfrac{(3.025 \text{ g})(100\%)}{286 \text{ mL}} = \mathbf{1.06 \text{ SDI}}$

ORGANIC LOADING RATE CALCULATIONS

Organic loading rate calculations tell the operator the amount of food entering the plant. These calculations are used for wastewater treatment ponds, rotating biological contactors, or trickling filters. See figures in Appendix E for the types of wastewater plants using these processes.

157. **A wastewater treatment pond receives a flow of 485,000 gpd. What is the organic loading rate in pounds of biological oxygen demand per day per acre (lb BOD$_5$/d/acre), if the pond has a surface area of 5.92 acre-ft and the influent BOD$_5$ concentration is 205 mg/L?**

First, convert gallons per day to mgd.

$$\text{Number of mgd} = \frac{485,000 \text{ gpd}}{1,000,000/\text{mil}} = 0.485 \text{ mgd}$$

Next, determine the pounds of BOD$_5$/d/acre-ft using a modified version of the "pounds" equation.

$$\textbf{Organic loading rate, lb BOD}_5\textbf{/d/acre-ft} = \frac{(\text{BOD}_5, \text{mg/L})\,(\text{Flow, mgd})\,(8.34 \text{ lb/gal})}{\text{Surface area of pond, acre-ft}}$$

$$\text{Organic loading rate, lb BOD}_5\text{/d/acre-ft} = \frac{(205 \text{ mg/L BOD}_5)\,(0.485 \text{ mgd})\,(8.34 \text{ lb/gal})}{5.92 \text{ acre-ft}}$$

Organic loading rate, lb BOD$_5$/d/acre-ft = 140.07 lb BOD$_5$/d/acre-ft, round to **140 lb BOD$_5$/d/acre-ft**

158. **What is the organic loading rate for a trickling filter in lb BOD$_5$/d/1,000 ft^3, given the following data?**

Trickling filter volume = 31,000 ft^3
Primary effluent flow = 4,020,000 gpd
BOD$_5$ = 152 mg/L

First, convert gallons per day to mgd.

$$\text{Number of mgd} = \frac{4,020,000 \text{ gpd}}{1,000,000/\text{mil}} = 4.02 \text{ mgd}$$

Next, factor out 1,000 ft^3 from the volume = (31)(1,000 ft^3)

Next, determine the pounds of BOD$_5$/d/1,000 ft^3 using a modified version of the "pounds" equation.

Organic loading rate, lb BOD$_5$/d/1,000 ft^3 $= \dfrac{\text{(BOD}_5\text{, mg/L)(Flow, mgd)(8.34 lb/gal)}}{\text{Volume of trickling filter, ft}^3/1,000 \text{ ft}^3}$

Organic loading rate, lb BOD$_5$/d/1,000 ft^3 $= \dfrac{(152 \text{ mg/L BOD}_5)\,(4.02 \text{ mgd})\,(8.34 \text{ lb/gal})}{(31)\,(1,000 \text{ ft}^3)}$

Organic loading rate, lb BOD$_5$/d/1,000 ft^3 = 164.39, round to **160 lb BOD$_5$/d/1,000 ft^3**

SUSPENDED SOLIDS LOADING CALCULATIONS

Operators use suspended solids loading for evaluating process control.

159. **What is the amount of suspended solids entering a trickling filter in lb/day, if the influent flow is 2.18 mgd and the amount of suspended solids (SS) is 325 mg/L?**

This problem uses the same equation as a dosage problem.

Equation: **Number of lb/day SS = (SS, mg/L)(Number of mgd)(8.34 lb/gal)**

Substitute values and solve.

Number of lb/day SS = (325 mg/L SS)(2.18 mgd)(8.34 lb/gal)

Number of lb/day SS = 5,908.89 lb/day, round to **5,910 lb/day SS**

160. **Determine the amount of suspended solids entering a trickling filter in mg/L, if the influent flow is 1.56 mgd and the suspended solids loading is 4,260 lb/day.**

Equation: **Number of lb/day SS = (SS, mg/L)(Number of mgd)(8.34 lb/gal)**

Rearrange the equation to solve for mg/L, SS.

$$\text{Number of mg/L SS} = \frac{\text{Number of lb/day SS}}{(\text{Number of mgd})(8.34\ \text{lb/gal})}$$

Substitute values and solve.

$$\text{Number of mg/L SS} = \frac{4,260\ \text{lb/day}}{(1.56\ \text{mgd})(8.34\ \text{lb/gal})} = 327.43\ \text{mg/L, round to } \textbf{327 mg/L SS}$$

BIOCHEMICAL OXYGEN DEMAND LOADING CALCULATIONS

Biochemical oxygen demand is the demand for oxygen made by bacteria as they decompose organic matter in wastewater or in the natural environment. This calculation sometimes is helpful in evaluating treatment pond processes. The BOD_5 is a 5-day test. See Figures 1 and 6 in Appendix E for two types of wastewater plants using a trickling filter.

161. **Calculate the biochemical oxygen demand (BOD_5) loading on a trickling filter in lb/day, if the influent flow into the trickling filter is 2.07 mgd and the influent BOD_5 is 168 mg/L.**

Equation: **Number of lb/day BOD_5 = (BOD_5, mg/L)(Number of mgd)(8.34 lb/gal)**

Substitute values and solve.

Number of lb/day BOD_5 = (168 mg/L BOD_5)(2.07 mgd)(8.34 lb/gal)

Number of lb/day BOD_5 = **2,900 lb/day BOD_5**

162. **Calculate the influent flow to a trickling filter in mgd, if the BOD$_5$ loading is 1,880 lb/day and the BOD$_5$ is 205 mg/L.**

Equation: **Number of lb/day BOD$_5$ = (BOD$_5$, mg/L)(Number of mgd)(8.34 lb/gal)**

Rearrange the equation to solve for mg/L.

$$\textbf{Number of mgd} = \frac{\text{Number of lb/day BOD}_5}{(\text{Number of mg/L BOD}_5)\,(8.34\ \text{lb/gal})}$$

Substitute values and solve.

$$\text{Number of mgd flow} = \frac{1{,}880\ \text{lb/day}}{(205\ \text{mg/L})\,(8.34\ \text{lb/gal})} = \textbf{1.10 mgd}$$

163. **What is the biochemical oxygen demand (BOD$_5$) loading on a trickling filter in lb/day, if the influent flow into the trickling filter is 1,250 gpm and the influent BOD$_5$ is 215 mg/L?**

First, convert gpm to mgd.

$$\text{Number of mgd} = \frac{(1{,}250\ \text{gpm})\,(1{,}440\ \text{min/day})}{1{,}000{,}000/\text{mil}} = 1.80\ \text{mgd}$$

Equation: **Number of lb/day BOD$_5$ = (BOD$_5$, mg/L)(Number of mgd)(8.34 lb/gal)**

Substitute values and solve.

Number of lb/day BOD$_5$ = (215 mg/L BOD$_5$)(1.80 mgd)(8.34 lb/gal)

Number of lb/day BOD$_5$ = 3,227.58 lb/day, round to **3,230 lb/day BOD$_5$**

SOLUBLE AND PARTICULATE BIOCHEMICAL OXYGEN DEMAND CALCULATIONS

BOD_5 measures the amount of organic matter that is present in water. Bacteria break down this organic matter by natural decomposition and in the process utilize oxygen. Thus the more organic matter present in the wastewater, the more demand for oxygen by the bacteria. Operators need to know how to do BOD_5 calculations because there are strict regulations for the amount of BOD_5 that can be discharged to a natural water body from the treated plant. The K value in these problems is the portion of suspended solids in the wastewater that are organic suspended solids. Domestic water usually has about 50 to 70% of the suspended solids as organic suspended solids, which is usually written in decimal form (0.5 to 0.7).

164. What is the approximate particulate biochemical oxygen demand (BOD_5) concentration for a wastewater, if the suspended solids (SS) are 208 mg/L and the K value for the plant is usually 0.575?

Equation: **Particulate BOD_5, mg/L = (SS, mg/L)(K value)**

Substitute values and solve.

Particulate BOD_5, mg/L = (208 mg/L SS)(0.575 K value)

Particulate BOD_5, mg/L = 119.6 mg/L, round to **120 mg/L Particulate BOD_5**

165. Given the following, calculate the estimated particulate BOD_5 in mg/L for the following wastewater.

Plant flow = 1,160 gpm
SS solids = 236 mg/L
K factor = 0.58

Equation: **Particulate BOD_5, mg/L = (SS, mg/L)(K factor)**

Substitute values and solve.

Particulate BOD_5, mg/L = (236 mg/L SS)(0.58 K factor)

Particulate BOD_5, mg/L = 136.88 mg/L, round to **140 mg/L Particulate BOD_5**

166. What is the soluble BOD_5 if the total BOD_5 is 248 mg/L, the K factor is 0.55, and the total suspended solids (SS) are 190 mg/L?

Equation: **Soluble BOD_5 = Total BOD_5 − (K factor)(Total SS)**

Substitute values and solve.

Soluble BOD_5 = 248 mg/L BOD_5 − (0.55 K factor)(190 mg/L SS)

Soluble BOD_5 = 248 mg/L − 104.5 mg/L

Soluble BOD_5 = 143.5 mg/L, round to **140 mg/L Soluble BOD_5**

167. What is the soluble BOD_5 if the suspended solids are 176 mg/L, the total BOD_5 is 208 mg/L, and the K factor is 0.55? Please note: Suspended solids are particulates.

Equation: **Total BOD_5 = (Particulate BOD_5)(K factor) + Soluble BOD_5**

Rearrange the equation to solve for soluble BOD_5.

Soluble BOD_5 = Total BOD_5 − (Particulate BOD_5)(K factor)

Substitute values and solve.

Soluble BOD_5 = 208 mg/L BOD_5 − (176 mg/L BOD_5)(0.55 K factor)

Soluble BOD_5 = 208 mg/L BOD_5 − 96.8 mg/L BOD_5

Soluble BOD_5 = 111.2 mg/L, round to **110 mg/L Soluble BOD_5**

168. Calculate the particulate BOD_5 given the following data:

K factor $= 0.63$
Total BOD_5 $= 231$ mg/L
Soluble BOD_5 $= 129$ mg/L

Equation: **Total BOD_5 = (Particulate BOD_5)(K factor) + Soluble BOD_5**

Rearrange the equation to solve for particulate BOD_5.

First, subtract soluble BOD_5 from both sides of the equation.

(Particulate BOD_5)(K factor) = Total BOD_5 − Soluble BOD_5

Next, divide both sides by the K factor.

$$\text{Particulate } BOD_5 = \frac{\text{Total } BOD_5 - \text{Soluble } BOD_5}{\text{K factor}}$$

Substitute values and solve.

$$\text{Particulate } BOD_5 = \frac{231 \text{ mg/L } BOD_5 - 129 \text{ mg/L } BOD_5}{0.63 \text{ factor}}$$

Particulate BOD_5 = 161.9 mg/L, round to **160 mg/L Particulate BOD_5**

CHEMICAL OXYGEN DEMAND LOADING CALCULATIONS

Chemical oxygen demand is a measure of the capacity of water to consume oxygen, when organic matter and the oxidation of inorganic matter are decomposed. Because it also measures the decomposition of inorganic matter such as nitrate and ammonia, it is only an indirect measure of the organic matter in water.

169. **Determine the amount of chemical oxygen demand (COD) entering an aeration tank in mg/L, if the influent flow is 2.66 mgd and the COD loading is 3,055 lb/day.**

Equation: **Number of lb/day COD = (COD, mg/L)(Number of mgd)(8.34 lb/gal)**

Rearrange the equation to solve for mg/L.

$$\text{Number of mg/L COD} = \frac{\text{Number of lb/day COD}}{(\text{Number of mgd})(8.34\,\text{lb/gal})}$$

Substitute values and solve.

$$\text{Number of mg/L COD} = \frac{3{,}055\,\text{lb/day}}{(2.66\,\text{mgd})(8.34\,\text{lb/gal})} = 137.7\,\text{mg/L, round to } \textbf{138 mg/L COD}$$

170. **What is the influent flow to an aeration tank in mgd, if the COD loading is 3,105 lb/day and the COD is 179 mg/L?**

Equation: **Number of lb/day COD = (COD, mg/L)(Number of mgd)(8.34 lb/gal)**

Rearrange the equation to solve for mg/L.

$$\text{Number of mgd} = \frac{\text{Number of lb/day COD}}{(\text{Number of mg/L COD})(8.34\,\text{lb/gal})}$$

Substitute values and solve.

$$\text{Number of mgd flow} = \frac{3{,}105\,\text{lb/day}}{(179\,\text{mg/L})(8.34\,\text{lb/gal})} = \textbf{2.08 mgd}$$

HYDRAULIC DIGESTION TIME CALCULATIONS

Hydraulic digestion time tells the operator how long the process will take to complete, and is thus used for planning purposes. See Figures 2, 4, 5, and 6 in Appendix E for four types of wastewater plants using a digester.

171. What is the hydraulic digestion time for a 49.5-ft diameter digester with a level of 9.2 ft and sludge flow of 8,230 gallons per day (gpd)?

First, determine the volume of the digester in ft^3.

Volume, ft^3 = (0.785)(Diameter)^2(Depth, ft)

Volume, ft^3 = (0.785)(49.5 ft)(49.5 ft)(9.2 ft) = 17,695.7 ft^3

Next, determine the number of gallons in the digester.

Know: 1 ft^3 = 7.48 gal/ft^3

Number of gal = (17,695.7 ft^3)(7.48 gal/ft^3) = 132,363.9 gal

Lastly, calculate the digestion time in days.

Equation: **Digestion time, days** $= \dfrac{\text{Number of gallons}}{\text{Influent sludge flow, gal/day}}$

Substitute values and solve.

Digestion time, days $= \dfrac{132,363.9 \text{ gal}}{8,230 \text{ gal/day}}$ = 16.08 days, round to **16 days**

172. **What is the hydraulic digestion time for a 32.2-ft diameter digester with a level of 9.5 ft and sludge flow of 5,880 gallons per day (gpd)?**

Equation: **Digestion time, days** $= \dfrac{(0.785)\,(\text{Diameter})^2(\text{Depth, ft})\,(7.48\,\text{gal/ft}^3)}{\text{Influent sludge flow, gal/day}}$

Substitute values and solve.

Digestion time, days $= \dfrac{(0.785)\,(32.2\,\text{ft})\,(32.2\,\text{ft})\,(9.5\,\text{ft})\,(7.48\,\text{gal/ft}^3)}{5,880\,\text{gpd}}$

Digestion time, days $=$ **9.8 days**

DIGESTER LOADING RATE CALCULATIONS

Digester loading rate calculations tell the operator how much volatile solids are stabilized per cubic foot of digester space. It is used for evaluating process control.

173. **Calculate the loading rate on a digester in pounds of volatile solids added (VSA) per day per cubic foot (lb VSA/d/ft³), if the volume of the digester is 28,200 ft³ and the digester has an influent of 9,350 lb/day of volatile solids.**

Equation: **Digester loading rate, lb VSA/d/ft³** $= \dfrac{\text{lb VSA}}{\text{Volume of digester, ft}^3}$

Digester loading rate, lb VSA/d/ft³ $= \dfrac{9,350\,\text{lb VSA}}{28,200,\,\text{ft}^3} =$ **0.33 lb VSA/d/ft³**

174. **Determine the loading rate on a digester in pounds of volatile solids added (VSA) per day per cubic foot (lb VSA/d/ft³), if the volume of sludge in the digester is 187,750 gal and the digester has an influent of 7,110 lb/day of volatile solids.**

First, convert the volume of the digester from gallons to cubic feet.

$$\text{Number of ft}^3, \text{digester} = \frac{187,750 \text{ gal}}{7.48 \text{ gal/ft}^3} = 25,100 \text{ ft}^3$$

Next, using the following equation, determine the loading rate on the digester.

Equation: **Digester loading rate, lb VSA/d/ft³** $= \dfrac{\text{lb VSA}}{\text{Volume of digester, ft}^3}$

$$\text{Digester loading rate, lb VSA/d/ft}^3 = \frac{7,110 \text{ lb VSA}}{25,100, \text{ft}^3} = \textbf{0.28 lb VSA/d/ft}^3$$

MEAN CELL RESIDENCE TIME (SOLIDS RETENTION TIME) CALCULATIONS

The mean cell residence time (MCRT) is the average time the activated-sludge solids are in an activated biosolids system. The MCRT is an important design and operating parameter for operators to use in the activated-sludge process and is normally expressed in days. This calculation is used for operational process control. See Figure 2 in Appendix E for one type of wastewater plant using the activated sludge process.

175. **Given the following data, calculate the mean cell residence time (MCRT) for this activated sludge system:**

Aeration tank plus secondary clarifier volume = 0.465 mil gal
Mixed liquor suspended solids (MLSS) = 3,720 mg/L
Suspended solids (SS) wasted = 1,820 lb/day
Secondary effluent SS = 385 lb/day

Equation: **MCRT, days** $= \dfrac{(\text{MLSS, mg/L})(\text{mil gal})(8.34 \text{ lb/day})}{\text{SS wasted, lb/day} + \text{SS lb/day}}$

Substitute values and solve.

$$\text{MCRT, days} = \frac{(3,720 \text{ mg/L})(0.465 \text{ mil gal})(8.34 \text{ lb/day})}{1,820 \text{ lb/day SS} + 385 \text{ lb/day SS}}$$

$$\text{MCRT, days} = \frac{(3,720 \text{ mg/L})(0.465 \text{ mil gal})(8.34 \text{ lb/day})}{2,205 \text{ lb/day SS}} = \textbf{6.54 days}$$

176. Given the following data, calculate the mean cell residence time (MCRT) for this activated sludge system:

Aeration tank volume = 0.62 mil gal
Secondary clarifier volume = 0.15 mil gal
Mixed liquor suspended solids (MLSS) = 3,430 mg/L
Suspended solids (SS) wasted = 2,180 lb/day
Secondary effluent SS = 320 lb/day

First, add the aerator and final clarifier volumes.

Total volume, mil gal = 0.62 mil gal + 0.15 mil gal = 0.77 mil gal

Equation: **MCRT, days** $= \dfrac{(\text{MLSS, mg/L})(\text{mil gal})(8.34 \text{ lb/day})}{\text{SS wasted, lb/day} + \text{SS lb/day}}$

Substitute values and solve.

MCRT, days $= \dfrac{(3,430 \text{ mg/L})(0.77 \text{ mil gal})(8.34 \text{ lb/day})}{2,180 \text{ lb/day SS} + 320 \text{ lb/day SS}}$

MCRT, days $= \dfrac{(3,430 \text{ mg/L})(0.77 \text{ mil gal})(8.34 \text{ lb/day})}{2,500 \text{ lb/day SS}} = 8.81$ days, round to **8.8 days**

DIGESTER VOLATILE SOLIDS LOADING RATIO CALCULATIONS

This calculation compares the volatile solids added to the volatile solids in the digester. It is used for evaluating process control.

177. If a digester has 24,200 lb of volatile solids (VS) and 2,750 lb/day are pumped into it, what is the ratio of volatile solids loading on the digester?

Use the following equation to solve for digester volatile solid ratio.

Equation: **Digester VS ratio** $= \dfrac{\text{VS added lb/day}}{\text{lb VS in digester}}$

Digester VS ratio $= \dfrac{2,750 \text{ lb/day}}{24,200 \text{ lb}} = $ **0.114 VS ratio**

178. **Given the following data, calculate the volatile solids (VS) loading ratio on a digester:**

Sludge weight in digester = 130,500 lb
VS loading = 1,255 lb/day
Total solids (TS) percentage = 4.89%
VS percentage = 66.2%

Use expanded equation with percentages:

$$\text{Digester VS ratio} = \frac{\text{VS added lb/day}}{(\text{lb VS in digester})(\text{TS \%}/100\%)(\text{VS \%}/100\%)}$$

$$\text{Digester VS ratio} = \frac{1,255 \text{ lb/day, VS}}{(130,500 \text{ lb VS})(4.89\%/100\% \text{ TS})(66.2\%/100\% \text{ VS})}$$

$$\text{Digester VS ratio} = \frac{1,255 \text{ lb/day, VS}}{(130,500 \text{ lb VS})(0.0489 \text{ TS})(0.662 \text{ VS})}$$

Digester VS ratio = **0.297 VS ratio**

179. **If the ratio of volatile solids (VS) added to volatile solids already in a digester is 0.08 and the amount of VS already in the digester is 29,500 lb, what is the amount in lb/day that must have been added?**

Equation: **Digester VS ratio** $= \dfrac{\text{VS added lb/day}}{\text{lb VS in digester}}$

Rearrange the problem to solve for volatile solids added.

VS added lb/day = (Digester VS ratio)(lb VS in digester)

Substitute values and solve.

VS added lb/day = (0.08 VS ratio)(29,500 lb VS in digester) = **2,360 lb/day VS added**

DIGESTER GAS PRODUCTION PROBLEMS

Operators calculate the amount of gases produced per pound of volatile solids destroyed to determine the effectiveness of the digestion process. Also, it is important to know the gas production because in some cases it is used as a fuel for other plant processes.

180. If a digester produces 8,420 ft³/day of gas and the amount of volatile solids destroyed are 725 lb/day, what is the amount of gas produced in ft³ per lb of volatile solids (VS) destroyed?

Equation: **Gas produced, ft³/lb VS destroyed** $= \dfrac{\text{Gas production, ft}^3/\text{day}}{\text{VS destroyed, lb/day}}$

Gas produced, ft³/lb VS destroyed $= \dfrac{8,420 \text{ ft}^3/\text{day}}{725 \text{ lb/day}} =$ **11.6 ft³/lb of VS destroyed**

181. Given the following data, calculate the gas produced by a digester in ft³ per pound of volatile solids (VS) destroyed:

Digester gas production = 5,770 ft³/day
Volatile solids destroyed = 435 lb/day

Equation: **Gas produced, ft³/lb VS destroyed** $= \dfrac{\text{Gas production, ft}^3/\text{day}}{\text{VS destroyed, lb/day}}$

Gas produced, ft³/lb VS destroyed $= \dfrac{5,770 \text{ ft}^3/\text{day}}{435 \text{ lb/day}} =$ **13.3 ft³/lb of VS destroyed**

182. What must have been the gas production by a digester in ft³/day, given the following data?

Volatile solids destroyed = 512 lb/day
Gas produced in ft³/lb VS destroyed = 12.6 ft³/lb

Equation: **Gas produced, ft³/lb VS destroyed** $= \dfrac{\text{Gas production, ft}^3/\text{day}}{\text{VS destroyed, lb/day}}$

Rearrange to solve for gas production in ft³/day.

Gas production, ft³/day = (Gas produced, ft³/lb VS destroyed)(VS destroyed, lb/day)

Substitute values and solve.

Gas production, ft³/day = (12.6 ft³/lb)(512 lb/day) = 6,451.2 ft³/day, round to **6,450 ft³/day**

VOLATILE ACIDS–TO–ALKALINITY RATIO PROBLEMS

The first phase of anaerobic digestion is acid fermentation, which is dependent on new volatile solids entering the digester. The second stage is methane fermentation. These two processes need to be in delicate balance with each other for the anaerobic digestion process to proceed properly. Different treatment plants have different ratios, but typically the ratio is less than 0.1.

183. Determine the ratio of volatile acids to alkalinity, if the alkalinity in an anaerobic digester is 2,054 mg/L and the volatile acid concentration of the sludge is 181 mg/L.

Equation: **Ratio = Volatile acids/Alkalinity**

Substitute values and solve.

$$\text{Ratio} = \frac{181 \text{ mg/L}}{2,054 \text{ mg/L}} = \textbf{0.0881}$$

184. What must have been the volatile acid concentration in an anaerobic digester, if the alkalinity was 1,942 mg/L and the ratio of volatile acids to alkalinity was 0.0849?

Equation: **Volatile acids = (Alkalinity)(Ratio)**

Substitute values and solve.

Volatile acids, mg/L = (1,942 mg/L)(0.0849) = 164.88 mg/L, round to **165 mg/L**

LIME NEUTRALIZATION PROBLEMS

When the sludge in an anaerobic digester becomes acidic it is called a sour digester. A sour digester occurs when the volatile acid–to–alkalinity ratio increases above 0.8. It is not always possible to wait for a digester to naturally correct itself because of the digester's capacity or time constraints. Under these circumstances, it is necessary to neutralize the acid conditions in the digester with lime. The following problems show how operators calculate the appropriate dosage of lime. The lime dosage is based on the amount of volatile acids in the sludge and is in a 1-to-1 ratio, that is, 1 mg/L of lime will neutralize 1 mg/L of volatile acid.

185. Given the following data, calculate the amount of lime in lb that is needed to neutralize a sour digester:

Digester volume = 198,500 gallons
Volatile acids = 1,961 mg/L

Know: 1 mg/L of lime will neutralize 1 mg/L of volatile acids

First, convert the digester's volume from gallons to million gallons.

$$\text{Number of mil gal} = \frac{198,500 \text{ gal}}{1,000,000/\text{mil}} = 0.1985 \text{ mil gal}$$

Equation: **Lime, lb = (Volatile acids, mg/L)(mil gal)(8.34 lb/gal)**

Substitute values and solve.

Lime, lb = (1,961 mg/L)(0.1985 mil gal)(8.34 lb/gal) = **3,246 lb of Lime**

186. If a sour digester has a volume of 225,000 gallons and a volatile acid concentration of 2,377 mg/L, how many pounds of lime will it take to neutralize the volatile acids?

Know: 1 mg/L of lime will neutralize 1 mg/L of volatile acids

First, convert the digester's volume from gallons to million gallons.

$$\text{Number of mil gal} = \frac{225,000 \text{ gal}}{1,000,000/\text{mil}} = 0.225 \text{ mil gal}$$

Equation: **Lime, lb = (Volatile acids, mg/L)(mil gal)(8.34 lb/gal)**

Substitute values and solve.

Lime, lb = (2,377 mg/L)(0.225 mil gal)(8.34 lb/gal) = **4,460 lb of Lime**

187. What must have been the concentration of volatile acids in mg/L for a sour digester with a volume of 254,000 gallons, if the number of pounds of lime to neutralize the volatile acids was 3,888 lb?

Know: 1 mg/L of lime will neutralize 1 mg/L of volatile acids

First, convert the digester's volume from gallons to million gallons.

$$\text{Number of mil gal} = \frac{254,000 \text{ gal}}{1,000,000/\text{mil}} = 0.254 \text{ mil gal}$$

Equation: **Number of lb = (Volatile acids, mg/L)(mil gal)(8.34 lb/gal)**

Rearrange the equation to solve for the concentration of volatile acids.

$$\text{Volatile acids, mg/L} = \frac{\text{Number of lb, lime}}{(\text{mil gal})\,(8.34 \text{ lb/gal})}$$

Substitute values and solve.

$$\text{Volatile acids, mg/L} = \frac{3,888 \text{ lb, lime}}{(0.254 \text{ mil gal})\,(8.34 \text{ lb/gal})}$$

Volatile acids, mg/L = 1,835.38 mg/L, round to **1,840 mg/L Volatile acids**

POPULATION LOADING CALCULATIONS

These calculations are used for wastewater treatment ponds. They are based on the number of people per acre of pond, and it is a helpful tool in evaluating process control of ponds.

188. Calculate the population loading in people per acre on three ponds, if the three ponds have a total area of 12.7 acres and they serve 23,500 people.

$$\textbf{Population loading, people/acre} = \frac{\text{Number of people served}}{\text{Area of pond(s), acres}}$$

Substitute values and solve.

$$\text{Population loading, people/acre} = \frac{23,500 \text{ people served}}{12.7 \text{ acres}} = \textbf{1,850 people/acre}$$

189. Given the following data, calculate the population loading in people per acre on two ponds.

Number one pond = 3.75 acres
Number two pond = 6.52 acres
Population served = 11,045 people

First, add the area in acres for each pond to get the total acres.

Total area of ponds, acres = 3.75 acres + 6.52 acres = 10.27 acres

Next, using the following equation determine the population loading.

$$\textbf{Population loading, people/acre} = \frac{\text{Number of people served}}{\text{Area of pond(s), acres}}$$

Substitute values and solve.

$$\text{Population loading, people/acre} = \frac{11,045 \text{ people served}}{10.27 \text{ acres}}$$

Population loading, people/acre = 1,075 people/acre, round to **1,080 people/acre**

POPULATION EQUIVALENT CALCULATIONS

Wastewater discharge from industries or commercial sources usually has a higher organic content than domestic wastewaters. Operators use population equivalent calculations to compare domestic wastewater to wastewater from these former sources. This is important in determining the loading that will be placed on a wastewater system when a new industry wants to connect to a system. What is needed is the flow from this industry in mgd and the BOD_5 concentration in mg/L. Domestic wastewater systems usually contain a range of 0.17 to 0.20 pounds of BOD_5 per day, which the wastewater plant should have already determined. Also, population equivalent calculations are required for designing proper size wastewater treatment plants, pump stations, and pipe sizes, because the volumetric flow that is expected to be treated and pumped needs to be estimated.

190. **Given the following data, calculate the population equivalent for this particular wastewater flow.**

Wastewater flow = 2.77 mgd
BOD_5/person = 0.21 lb/day of BOD_5/person
BOD_5 concentration in the wastewater = 2,460 mg/L

Use the following equation to solve this problem.

$$\textbf{Number of people} = \frac{(BOD_5, \text{mg/L}) (\text{mgd}) (8.34 \text{ lb/gal})}{\text{lb/day of } BOD_5/\text{person}}$$

$$\text{Number of people} = \frac{(2,460 \text{ mg/L } BOD_5) (2.77 \text{ mgd}) (8.34 \text{ lb/gal})}{0.21 \text{ lb/day}}$$

Number of people = 270,621 people, round to **270,000 people**

191. **A wastewater treatment plant has an average flow of 1,375,000 gallons. If the BOD$_5$ is 2,050 mg/L and the average BOD$_5$ per person is 0.25 lb/day, what is the population equivalent that this plant is currently treating?**

First, convert the flow in gallons to mgd.

$$\text{Number of mgd} = \frac{1,375,000 \text{ gal}}{1,000,000/\text{mil}} = 1.375 \text{ mgd}$$

Next, use the following equation to solve this problem.

$$\textbf{Number of people} = \frac{(\text{BOD}_5, \text{mg/L})(\text{mgd})(8.34 \text{ lb/gal})}{\text{lb/day of BOD}_5/\text{person}}$$

$$\text{Number of people} = \frac{(2,050 \text{ mg/L BOD}_5)(1.375 \text{ mgd})(8.34 \text{ lb/gal})}{0.25 \text{ lb/day}}$$

Number of people = 94,034 people, round to **94,000 people**

SOLIDS UNDER AERATION

Solids under aeration calculations are used by operators for evaluating process control.

192. **Calculate the number of pounds of suspended solids contained in an aeration tank, if the tank contains 263,000 gallons and the concentration of suspended solids is 1,988 mg/L.**

First, convert gallons to million gallons.

$$\text{Number of mil gal} = \frac{263,000 \text{ gal}}{1,000,000/\text{mil}} = 0.263 \text{ mil gal}$$

Next, determine the pounds of suspended solids using the "pounds" equation.

Equation: **Number of lb, solids = (SS, mg/L)(Number of mil gal)(8.34 lb/gal)**

Substitute values and solve.

Number of lb SS = (1,988 mg/L SS)(0.263 mil gal)(8.34 lb/gal)

Number of lb SS = 4,360.52 lb, round to **4,360 lb SS**

193. **The capacity of an aeration tank is 325,000 gallons. How many pounds of MLSS are being aerated, if the concentration of MLSS is 2,405 mg/L?**

First, determine how many gallons are in the aeration tank.

Next, convert gallons to million gallons.

$$\text{Number of mil gal} = \frac{325,000 \text{ gal}}{1,000,000/\text{mil}} = 0.325 \text{ mil gal}$$

Next, determine the pounds of MLSS under aeration using the "pounds" equation.

Equation: **Number of lb = (MLSS, mg/L)(Number of mil gal)(8.34 lb/gal)**

Substitute values and solve.

Number of lb MLSS = (2,405 mg/L MLSS)(0.325 mgd)(8.34 lb/gal)

Number of lb MLSS = 6,518.75 lb, round to **6,520 lb MLSS**

SUSPENDED SOLIDS REMOVAL

The suspended solids removal calculations are used by operators as a sign for the efficiency of the treatment process in question. Typically, the suspended solids removed from wastewater systems ranges from 100 to 350 mg/L.

194. **What quantity of suspended solids in lb/day was removed from a primary clarifier that had an influent flow of 4.33 mgd and 92 mg/L of suspended solids (SS)?**

Equation: **SS removed, lb/day = (SS, mg/L)(Number of mgd)(8.34 lb/gal)**

Substitute values and solve.

Number of lb/day SS removed = (92 mg/L SS)(4.33 mgd)(8.34 lb/gal)

Number of lb/day SS removed = 3,322.32 lb/day, round to **3,300 lb/day SS removed**

195. If a wastewater treatment plant's clarifier had a flow of 2.77 mgd and a removal of 2,600 lb/day of suspended solids, what must have been the average influent concentration of suspended solids in mg/L?

Equation: **SS removed, lb/day = (SS, mg/L)(Number of mgd)(8.34 lb/gal)**

Rearrange the equation, then substitute values and solve.

$$\textbf{SS removed, mg/L} = \frac{(\text{SS removed, lb/day})}{(\text{Number of mgd})(8.34\ \text{lb/gal})}$$

$$\text{SS removed, mg/L} = \frac{(2,600\ \text{lb/day})}{(2.77\ \text{mgd})(8.34\ \text{lb/gal})} = 112.54\ \text{mg/L, round to } \textbf{110 mg/L SS}$$
removed

BIOCHEMICAL OXYGEN DEMAND REMOVAL CALCULATIONS

The biochemical oxygen demand (BOD_5) removal calculations are used to inform operators about the efficiency of the treatment process for a pond or trickling filter. The BOD_5 is an empirical test that informs the operator on the relative oxygen requirements of a wastewater, and is an indicator of how much food is in the wastewater. The BOD_5 is a 5-day test.

196. Given the following data, determine the amount in lb/day removal of biochemical oxygen demand (BOD_5) from a trickling filter.

Plant influent flow = 2.49 mgd
Influent BOD$_5$ concentration = 342 mg/L
Effluent BOD$_5$ concentration = 196 mg/L

First, determine the amount of BOD_5 removed in mg/L by subtracting the influent BOD_5 from the effluent BOD_5.

BOD_5 removed, mg/L = (Influent BOD_5, mg/L − Effluent BOD_5, mg/L)

BOD_5 removed, mg/L = 342 mg/L − 196 mg/L = 146 mg/L of BOD_5 removed

Next, solve the amount of BOD_5 removed in lb/day by using the "pounds" equation.

Equation: **BOD_5 removed, lb/day = (BOD_5, mg/L)(Number of mgd)(8.34 lb/gal)**

Substitute values and solve.

BOD_5 removed, lb/day = (146 mg/L BOD_5)(2.49 mgd)(8.34 lb/gal)

BOD_5 removed, lb/day = 3,031.92 lb/day, round to **3,030 lb/day BOD_5 removed**

197. What must have been the daily flow to a trickling filter in mgd, if the BOD_5 removed was 3,730 lb/day, the influent BOD_5 was 290 mg/L, and the effluent BOD_5 was 78 mg/L?

First, the amount of BOD_5 removed must still be determined by subtracting the influent BOD_5 from the effluent BOD_5.

BOD_5 removed, mg/L = 290 mg/L − 78 mg/L = 212 mg/L of BOD_5 removed

Next, solve the amount of BOD_5 removed in lb/day by using the "pounds" equation.

Equation: **Number of lb/day = (BOD_5, mg/L)(Number of mgd)(8.34 lb/gal)**

Rearrange the equation, then substitute values and solve.

$$\textbf{Number of mgd} = \frac{\text{Number of lb/day}}{(BOD_5 \text{ removed, mg/L})(8.34 \text{ lb/gal})}$$

Substitute values and solve.

$$\text{Number of mgd} = \frac{3,730 \text{ lb/day}}{(212 \text{ mg/L } BOD_5 \text{ removed})(8.34 \text{ lb/gal})} = \textbf{2.1 mgd}$$

FOOD-TO-MICROORGANISM RATIO CALCULATIONS

A properly operated activated sludge process has a balance between the food entering the system and the microorganisms in the aeration tank. The best ratio varies because it depends on the activated sludge process and the characteristics of the wastewater being treated. The calculation is a measure of the pounds of food coming in divided by the pounds of microorganisms present. The ratio is a process control number because it helps the operator determine the proper number of microorganisms for the system in question.

198. **Given the following data on an aeration tank, calculate the current food-to-microorganism (F/M) ratio.**

Primary effluent flow = 2.85 mgd
Volume of aeration tank = 207,000 gallons
Mixed liquor volatile suspended solids (MLVSS) = 2,450 mg/L
BOD_5 = 215 mg/L

First, convert the volume of wastewater in the tank to million gallons.

$$\text{Number of mil gal} = \frac{207,000 \text{ gal}}{1,000,000/\text{mil}} = 0.207 \text{ mil gal}$$

Next, write the equation: $\mathbf{F/M} = \dfrac{(BOD_5, \text{mg/L})(\text{Flow, mgd})(8.34 \text{ lb/gal})}{(\text{mg/L MLVSS})(\text{Volume in tank, mil gal})(8.34 \text{ lb/gal})}$

The 8.34 lb/gal cancels, leaving the following equation:

$$F/M = \frac{(BOD_5, \text{mg/L})(\text{Flow, mgd})}{(\text{mg/L MLVSS})(\text{Volume of tank, mil gal})}$$

Substitute values and solve.

$$F/M = \frac{(215 \text{ mg/L } BOD_5)(2.85 \text{ mgd})}{(2,450 \text{ mg/L MLVSS})(0.207 \text{ mil gal})} = \mathbf{1.21 \text{ F/M ratio}}$$

199. **What is the food-to-microorganism (F/M) ratio for an aeration tank with a volume of 325,000 gallons, if the primary effluent flow is 4.20 mgd, the MLVSS is 3,015 mg/L, and the BOD$_5$ is 241 mg/L?**

First, convert the volume of wastewater in the tank to million gallons.

Number of mil gal = 325,000 gal/1,000,000/mil = 0.325 mil gal

Next, write the equation:

$$\text{F/M} = \frac{(\text{BOD}_5,\ \text{mg/L})\,(\text{Flow, mgd})}{(\text{mg/L MLVSS})\,(\text{Volume of tank, mil gal})}$$

Substitute values and solve.

$$\text{F/M} = \frac{(241\ \text{mg/L BOD}_5)\,(4.20\ \text{mgd})}{(3,015\ \text{mg/L MLVSS})\,(0.325\ \text{mil gal})} = \textbf{1.03 F/M ratio}$$

SEED SLUDGE PROBLEMS

This calculation is required for determining how much seed sludge in gallons to use for starting a new digester.

200. **A digester has a capacity of 252,000 gallons. If the seed sludge required is 15% of the digester capacity, how many gallons of seed sludge will be needed?**

Equation: **Seed sludge, gal** $= \dfrac{(\text{Capacity of digester})\,(\text{Percent seed sludge required})}{100\%}$

Seed sludge, gal $= \dfrac{(252,000\ \text{gallons})\,(15\%)}{100\%} = \textbf{37,800 gal of seed sludge}$

201. **Given the following data, determine the seed sludge required in gallons:**

Digester has a radius of 27.5 ft
Liquid level in digester is 17.3 ft
Requires 17% seed sludge

First, determine the number of gallons in the digester.

Volume, gal $= \pi r^2$(Depth, ft)(7.48 gal/ft^3)

Volume, gal $= (3.14)(27.5 \text{ ft})(27.5 \text{ ft})(17.3 \text{ ft})(7.48 \text{ gal/ft}^3) = 307,286 \text{ gal}$

Next, use the following equation.

Seed sludge, gal $= \dfrac{\text{(Capacity of digester)(Percent seed sludge required)}}{100\%}$

Seed sludge, gal $= \dfrac{(307,286 \text{ gallons})(17\%)}{100\%}$

Seed sludge, gal $= 52,239$ gal, round to **52,000 gal of Seed sludge**

GRAVITY THICKENER SOLID LOADING PROBLEMS

Gravity thickeners use large tanks that separate suspended solids and mineral matter from the liquid by gravity. The gravity thickener concentrates the sludge to reduce the load on processes that follow—conditioning, dewatering, and digestion—and produces a clear liquid, which is decanted. Flocculants are used to speed up the settling process. Operators can calculate the solids loading in lb/d/ft^2 or the hydraulic loading in gal/day/ft^2. The hydraulic loading calculation is used by operators to determine if the process is being overloaded or underloaded. See Figures 4, 5, and 6 in Appendix E for three types of wastewater plants using the thickening process.

202. **Given the following data, determine the hydraulic loading on a gravity thickener in gal/d/ft^2:**

Gravity thickener $=$ 50.3 ft in diameter
Influent flow $=$ 38.4 gpm
Percent solids $=$ 3.1%

First, determine the area of the gravity thickener.

Know: **Area** $= \pi r^2$, where $\pi = 3.14$

Radius (r) = Diameter/2 = 50.3 ft/2 = 25.15 ft

Area $= (\pi)(25.15 \text{ ft})(25.15 \text{ ft}) = 3.14(632.5225 \text{ ft}^2) = 1,986.12 \text{ ft}^2$

Equation: **Hydraulic loading, gal/d/ft²** $= \dfrac{(\text{Flow, gpm})(1,440 \text{ min/day})(\text{Percent solids})}{(\text{Gravity thickener area})(100\%)}$

Substitute values and solve.

Hydraulic loading, gal/d/ft² $= \dfrac{(38.4 \text{ gpm})(1,440 \text{ min/day})(3.1\%)}{(1,986.12 \text{ ft}^2)(100\%)}$

Hydraulic loading, gal/d/ft² $= \mathbf{0.86 \text{ gal/d/ft}^2}$

203. A gravity thickener with a radius of 32.4 ft receives a flow of 0.0854 mgd. If the percent solids are 4.40%, what are the solids loading on the gravity thickener in lb/d/ft²?

First, convert the mgd to gpd.

Number of gpd = (0.0854 mgd)(1,000,000/mil) = 85,400 gpd

Know: Area of gravity thickener $= \pi r^2$ where $\pi = 3.14$

Equation: **Solids loading, lb/d/ft²** $= \dfrac{(\text{Flow, gpd})(8.34 \text{ lb/gal})(\text{Percent solids})}{(\text{Gravity thickener area})(100\%)}$

Substitute values and solve.

Solids loading, lb/d/ft² $= \dfrac{(85,400 \text{ gpd})(8.34 \text{ lb/gal})(4.40\%)}{(3.14)(32.4 \text{ ft})(32.4 \text{ ft})(100\%)}$

Solids loading, lb/d/ft² $= \mathbf{9.51 \text{ lb/d/ft}^2}$

DISSOLVED AIR FLOTATION: THICKENER SOLIDS LOADING PROBLEMS

The dissolved air flotation technique is used to thicken sludge. These types of calculations are used for evaluating process control.

204. **Given the following data, calculate the solids loading in lb/d/ft² on a dissolved air flotation (DAF) thickener unit:**

Area of DAF = 1,550 ft²
Waste-activated sludge (WAS) = 9,037 mg/L
Sludge flow = 145 gpm
Sludge weight = 8.84 lb/gal

First, convert gpm to mgd.

$$\text{Number of gpd} = \frac{(145\ \text{gpm})\,(1{,}440\ \text{min}/\text{day})}{1{,}000{,}000/\text{mil}} = 0.2088\ \text{mgd}$$

Equation: **Solids loading, lb/d/ft²** $= \dfrac{(\text{WAS, mg/L})\,(\text{mgd})\,(\text{lb/gal, Sludge})}{\text{DAF area, ft}^2}$

Solids loading, lb/d/ft² $= \dfrac{(9{,}037\ \text{mg/L, WAS})\,(0.2088\ \text{mgd})\,(8.84\ \text{lb/gal, Sludge})}{1{,}550\ \text{ft}^2\ \text{DAF}}$

Solids loading, lb/d/ft² = 10.76 lb/d/ft², round to **10.8 lb/d/ft²**

205. **What is the solids loading for a dissolved air flotation (DAF) unit in lb/hr/ft² that is 59.8 ft by 17.5 ft, with a sludge flow of 0.208 mgd, a waste-activated sludge (WAS) concentration of 7,548 mg/L, and the sludge weighs 8.79 lb/gal?**

First, determine the area of the DAF unit in ft².

DAF area, ft² = (59.8 ft)(17.5 ft) = 1,046.5 ft²

Next, calculate the solids loading using the following equation.

Equation: **Solids loading, lb/hr/ft²** $= \dfrac{(\text{WAS, mg/L})\,(\text{mgd})\,(\text{lb/gal, Sludge})}{(\text{DAF area, ft}^2)\,(24\ \text{hr/day})}$

Solids loading, lb/hr/ft² $= \dfrac{(7{,}548 \text{ mg/L, WAS})\,(0.208 \text{ mgd})\,(8.79 \text{ lb/gal, Sludge})}{(1{,}046.5 \text{ ft}^2 \text{ DAF})\,(24 \text{ hr/day})}$

Solids loading, lb/hr/ft² = 0.5495 lb/hr/ft², round to **0.550 lb/hr/ft²**

DISSOLVED AIR FLOTATION: AIR-TO-SOLIDS RATIO CALCULATIONS

Air-to-solids ratio calculations are used to determine the efficiency of the process, as the air flotation thickener and the solids in the system must be in balance. Typically the ratio ranges from 0.01 to 0.1.

206. **What is the air-to-solids ratio for a dissolved air flotation (DAF) unit that has an air flow rate of 7.8 ft³/min, a solids concentration of 0.65%, and a flow of 117 gpm?**

Know: Air = 0.0807 lb/ft³ at standard temperature, pressure, and average composition

Equation: **Air-to-solids ratio** $= \dfrac{(\text{Air flow, ft}^3/\text{min})\,(\text{Air, lb/ft}^3)}{(\text{gpm})\,(\text{Percent solids}/100\%)\,(8.34 \text{ lb/gal})}$

Substitute values and solve.

Air-to-solids ratio $= \dfrac{(7.8 \text{ ft}^3/\text{min})\,(100\%)\,(0.0807 \text{ lb/ft}^3)}{(117 \text{ gpm})\,(0.65\%)\,(8.34 \text{ lb/gal})} =$ **0.10 Air-to-solids ratio**

207. **Given the following data, determine the air-to-solids ratio for a DAF unit.**

DAF influent flow = 124 gpm
Air flow = 7.2 ft³/min
Solids concentration = 0.67%

Know: Air = 0.0807 lb/ft³ at standard temperature, pressure, and average composition

Equation: **Air-to-solids ratio** $= \dfrac{(\text{Air flow, ft}^3/\text{min})\,(\text{Air, lb/ft}^3)}{(\text{gpm})\,(\text{Percent solids}/100\%)\,(8.34 \text{ lb/gal})}$

Substitute values and solve.

Air-to-solids ratio $= \dfrac{(7.2 \text{ ft}^3/\text{min})\,(100\%)\,(0.0807 \text{ lb/ft}^3)}{(124 \text{ gpm})\,(0.67\%)\,(8.34 \text{ lb/gal})} =$ **0.084 Air-to-solids ratio**

DISSOLVED AIR FLOTATION: AIR RATE FLOW CALCULATIONS

Operators use air rate flow calculations for evaluating process control.

208. **If a DAF unit receives air at an average rate of 9.86 ft³/min, how many lb/day of air does it receive?**

Know: Air = 0.0807 lb/ft³ at standard temperature, pressure, and average composition

Equation: **Air, lb/day = (Air flow, ft³/min)(1,440 min/day)(0.0807 lb/ft³, Air)**

Substitute values and solve.

Air, lb/day = (9.86 ft³/min)(1,440 min/day)(0.0807 lb/ft³)

Air, lb/day = 1,145.81 lb/day, round to **1,150 lb/day of Air**

209. **If a DAF unit receives air at an average rate of 7.05 ft³/min, how many lb/hr of air does it receive?**

Know: Air = 0.0807 lb/ft³ at standard temperature, pressure, and average composition

Equation: **Air, lb/day = (Air flow, ft³/min)(60 min/hr)(0.0807 lb/ft³, Air)**

Substitute values and solve.

Air, lb/day = (7.05 ft³/min)(60 min/hr)(0.0807 lb/ft³)

Air, lb/day = 34.14 lb/hr, round to **34.1 lb/hr of Air**

CENTRIFUGE THICKENING PROBLEMS

Centrifuges are used to dewater sludge usually after applying gravity thickening. They apply forces that are a thousand times greater than gravity. Polymers may be applied to the influent of the centrifuge to facilitate solids thickening.

210. **What is the hydraulic loading on a unit in gal/day, if the disc centrifuge receives a waste-activated sludge flow of 29 gpm?**

Equation: **Hydraulic loading, gal/day = (Sludge flow, gpm)(1,440 min/day)**

Number of gal/day = (29 gpm)(1,440 min/day) = 41,760 gal/day, round to **42,000 gal/day**

211. **Given the following data, determine the removal efficiency of sludge solids on a disc centrifuge.**

Influent sludge solids = 6,385 mg/L
Effluent sludge solids = 525 mg/L

First, determine the amount of sludge solids removed.

Sludge solids removed = 6,385 mg/L − 525 mg/L = 5,860 mg/L

Now, determine the removal efficiency.

$$\textbf{Percent removal efficiency} = \frac{\text{(Solids removed, mg/L)}\,(100\%)}{\text{Influent solids, mg/L}}$$

$$\text{Percent removal efficiency} = \frac{(5,860\text{ mg/L})\,(100\%)}{6,385\text{ mg/L}} = 91.778\%,\text{ round to } \textbf{91.8\%}$$

SAND DRYING BED PROBLEMS

By knowing how thick the sludge was when applied and later measuring the thickness of the dried sludge, an operator can use these calculations to determine the efficiency of the drying bed process.

212. **A drying bed is 312 ft long and 40.5 ft wide. If 6.0 inches of sludge were applied to the drying bed, how many gallons of sludge were applied?**

First, convert 6.0 inches to feet.

Number of feet = 6.0 in./12 in./ft = 0.50 ft

Next, determine the volume in ft^3 sent to the drying bed.

Volume, ft^3 = (312 ft)(40.5 ft)(0.50 ft) = 6,318 ft^3

Lastly, calculate the volume in gallons sent to the sand drying beds.

Number of gal = (6,318 ft^3)(7.48 gal/ft^3) = 47,258.64 gal, round to **47,000 gal**

213. **A sand drying bed is 152 ft long and 39 ft wide. If a digester that is 49.5 ft in diameter is drawn down by 3.5 ft, how many cubic feet of sludge will be sent to the sand drying beds?**

Equation: **Volume, gal = (0.785)(Diameter, ft)^2(Depth, ft)**

Volume, gal = (0.785)(49.5 ft)(49.5 ft)(3.5 ft) = 6,732 ft^3, round to **6,700 ft^3**

Note: The size of the sand drying bed is unnecessary information, which you may encounter on a certification test.

DEWATERING CALCULATIONS

This section contains several types of dewatering problems. The more water removed from sludge, the less cost associated with further processing or disposal. The problems are important to the operator because they are helpful in evaluating process control or in informing the operator of process efficiency. See Figure 12 in Appendix E for one type of sludge process using dewatering.

214. **A feed sludge sample is collected from a belt filter press dewatering unit for laboratory testing to determine the total nonfilterable residue. If the feed sample contained a total residue of 28,500 mg/L and the filterable residue was 695 mg/L, what is the estimated total nonfilterable residue (total suspended solids [SS]) in mg/L?**

Equation:

Total nonfilterable residue, mg/L = Total residue, mg/L − Total filterable residue, mg/L

Substitute values and solve.

Total nonfilterable residue, mg/L = 28,500 mg/L − 695 mg/L

Total nonfilterable residue, mg/L = 27,805 mg/L, round to **27,800 mg/L Total SS**

215. **If the maximum feed rate to a belt filter press for an adequate cake is 1,500 lb/hr, how long in hours will it take to process 33,000 lb of sludge?**

$$\text{Number of hours} = \frac{33,000 \text{ lb}}{1,500 \text{ lb/hr}} = \textbf{22 hours}$$

216. **What is the sludge feed rate for a belt filter press to process 11,050 lb/day of sludge, if it operates only 9.00 hr/day?**

$$\text{Sludge feed rate, lb/hr} = \frac{11,050 \text{ lb/day}}{9.00 \text{ hr/day}} = 1,227.78 \text{ lb/hr, round to } \textbf{1,230 lb/hr}$$

217. A vacuum filter has a wet cake flow of 3,975 lb/hr and a filter area of 399.8 ft². Calculate the filter yield in lb/hr/ft², if the percent solids are 26%.

Equation: **Filter yield, lb/hr/ft²** $= \dfrac{(\text{Wet cake flow, lb/hr})(\text{Percent solids}/100\%)}{\text{Area, ft}^2}$

Substitute values and solve.

Filter yield, lb/hr/ft² $= \dfrac{(3{,}975 \text{ lb/hr})(26\%/100\%)}{399.8 \text{ ft}^2} = 2.585$ lb/hr/ft², round to **2.6 lb/hr/ft²**

218. Given the following data, calculate the filter loading in lb/day/ft² on a vacuum filter that has a surface area of 328.5 ft².

Digested biosolids flow rate = 55 gpm
Solids concentration = 3.6%

First, convert gpm to gal/day.

Number of gpd = (55 gpm)(1,440 min/day) = 79,200 gpd

Next, solve for filter loading.

Equation: **Vacuum filter loading, lb/day/ft²** $= \dfrac{(\text{Biosolids, gpd})(8.34 \text{ lb/gal})(\text{Percent solids})}{\text{Vacuum filter area, ft}^2}$

Vacuum filter loading, lb/day/ft² $= \dfrac{(79{,}200 \text{ gpd})(8.34 \text{ lb/gal})(3.6\%/100\%)}{328.5 \text{ ft}^2}$

Vacuum filter loading, lb/day/ft² = 72.39 lb/day/ft², round to **72 lb/day/ft²**

219. **Calculate the filter loading in lb/hr/ft² on a vacuum filter that has a surface area of 285 ft², a digested biosolids flow rate of 63 gpm, and a solids concentration of 4.12%.**

First, convert gpm to gal/hr.

Number of gal/hr = (63 gpm)(60 min/hr) = 3,780 gal/hr

Next, solve for filter loading.

Equation: **Vacuum filter loading, lb/hr/ft² =**
$$\frac{\text{(Biosolids, gal/hr)(8.34 lb/gal)(Percent solids)}}{\text{Vacuum filter area, ft}^2}$$

Vacuum filter loading, lb/hr/ft² = $\dfrac{(3{,}780 \text{ gal/hr})(8.34 \text{ lb/gal})(4.12\%/100\%)}{285 \text{ ft}^2}$

Vacuum filter loading, lb/hr/ft² = 4.557 lb/hr/ft², round to **4.6 lb/hr/ft²**

VACUUM FILTER YIELD AND OPERATING TIME CALCULATIONS

The vacuum filter yield calculation is used to measure performance. The operating time is the time required to process the solids and is used for planning purposes.

220. **A vacuum filter with a surface area of 245 ft² processes an average of 2,950 lb/day. If the solids recovery averages 93.8% and the vacuum filter yield averages 1.85 lb/hr/ft, how many hours per day will it take the vacuum filter to process these solids?**

Equation: **Filter yield, lb/hr/ft² =** $\dfrac{\dfrac{\text{(Solids, lb/day)(Percent recovery)}}{\text{(Filter operation, lb/day)(100\%)}}}{\text{Filter area ft}^2}$

Substitute values and solve.

1.85 lb/hr/ft² = $\dfrac{\dfrac{(2{,}950 \text{ lb/day})(93.8\%)}{\text{(Filter operation, hr/day)(100\%)}}}{245 \text{ ft}^2}$

Think of the above equation as:

$$1.85 \text{ lb/hr/ft}^2 = \frac{\dfrac{(2{,}950 \text{ lb/day})(93.8\%)}{(\text{Filter operation, hr/day})(100\%)}}{\dfrac{245 \text{ ft}^2}{1}}$$

The mathematical rule states you invert and multiply as follows:

$$1.85 \text{ lb/hr/ft}^2 = \frac{(2{,}950 \text{ lb/day})(93.8\%)}{(\text{Filter operation, hr/day})(100\%)} \times \frac{(1)}{245 \text{ ft}^2}$$

Simplify:

$$1.85 \text{ lb/hr/ft}^2 = \frac{(2{,}950 \text{ lb/day})(1)(93.8\%)}{(\text{Filter operation, hr/day})(245 \text{ ft}^2)(100\%)}$$

Rearrange the equation to solve for the unknown filter operation time.

$$\text{Filter operation, hr/day} = \frac{(2{,}950 \text{ lb/day})(1)(93.8\%)}{(1.85 \text{ lb/hr/ft}^2)(245 \text{ ft}^2)(100\%)} = \textbf{6.10 hr/day}$$

221. Given the following data, calculate the time in hr/day for a vacuum filter to process 3,560 lb/day:

Vacuum filter surface area = 224 ft²
Solids recovery averages 94.2%
Percent yield averages 1.94 lb/hr/ft

Equation: **Filter yield, lb/hr/ft²** $= \dfrac{(\text{Solids, lb/day})(\text{Percent recovery})}{\dfrac{(\text{Filter operation, lb/day})(100\%)}{\text{Filter area ft}^2}}$

Substitute values and solve.

$$1.94 \text{ lb/hr/ft}^2 = \frac{\dfrac{(3{,}560 \text{ lb/day})(94.2\%)}{(\text{Filter operation, hr/day})(100\%)}}{224 \text{ ft}^2}$$

Simplify:

$$1.94 \text{ lb/hr/ft}^2 = \frac{(3{,}560 \text{ lb/day})(1)(94.2\%)}{(\text{Filter operation, hr/day})(224 \text{ ft}^2)(100\%)}$$

Rearrange the equation to solve for the unknown, filter operation time.

$$\text{Filter operation, hr/day} = \frac{(3{,}560 \text{ lb/day})(1)(94.2\%)}{(1.94 \text{ lb/hr/ft}^2)(224 \text{ ft}^2)(100\%)} = \textbf{7.72 hr/day}$$

BIOSOLIDS VOLUME INDEX AND BIOSOLIDS DENSITY INDEX CALCULATIONS

Both of these calculations help determine the pumping rate of return biosolids and are a measure of the settling characteristics of the activated biosolids. They also exhibit how well the liquids/solids separation in the activated biosolids system is performing its role on the biological floc that has been generated. The goal is to produce a small volume of biosolids and thus reduce the pumping rate that is required for the solids to stay in circulation.

222. What is the biosolids volume index (BVI), given the following data?

Settleability test after 30 minutes = 283 mL of biosolids in 1-liter graduated cylinder
Aeration tank, mixed liquor suspended solids (MLSS) = 2,530 mg/L

Equation: $$\mathbf{BVI} = \frac{(\text{Settled biosolids, mL/L})\,(1,000\text{ mg/g})}{\text{MLSS, mg/L}}$$

Substitute values and solve.

$$\text{BVI} = \frac{(283\text{ mL/L})\,(1,000\text{ mg/g})}{2,530\text{ mg/L}}\ 111.86\text{ mL/g, which equates and rounds to }\mathbf{112\ BVI}$$

223. A settleability test result shows that 268 mL of activated biosolids settled in a 1-liter graduated cylinder. If the concentration of mixed liquor suspended solids (MLSS) is 2,720 mg/L, what is the biosolids density index (BDI)?

Equation: $$\mathbf{BDI} = \frac{(\text{MLSS, mg/L})\,(100)}{(\text{Settled biosolids, mL/L})\,(1,000\text{ mg/g})}$$

Substitute values and solve.

$$\text{BDI} = \frac{(2,720\text{ mg/L})\,(100)}{(268\text{ mL/L})\,(1,000\text{ mg/g})} = 1.01\text{ g/mL, which equates to }\mathbf{1.01\ BDI}$$

SETTLEABLE SOLIDS CALCULATIONS

These tests are performed on samples from either the clarifier's influent or effluent or from a sedimentation tank. They are used to determine the percent and thus the efficiency of settleable solids. Calculations based on these tests follow:

224. **Exactly 2,000 mL of activated sludge was collected in a graduated cylinder. What is the percent of settleable solids, if after exactly 30 minutes the sludge solids that settled totaled 456 mL?**

Equation: **Percent settleable solids** $= \dfrac{(\text{Settled sludge, mL})(100\%)}{\text{Sample size, mL}}$

Substitute values and solve.

Percent settleable solids $= \dfrac{(456 \text{ mL})(100\%)}{2,000 \text{ mL}} =$ **22.8% Settled solids**

225. **Given the following data, calculate the percent settleable solids:**

Activated sludge sample = 2,000 mL
Settling time is exactly = 30 minutes
Sludge solids in graduated cylinder = 308 mL

Equation: **Percent settleable solids** $= \dfrac{(\text{Settled sludge, mL})(100\%)}{\text{Sample size, mL}}$

Substitute values and solve.

Percent settleable solids $= \dfrac{(308 \text{ mL})(100\%)}{2,000.0 \text{ mL}} =$ **15.4% Settled solids**

CHEMISTRY AND LABORATORY PROBLEMS

Operators should have a thorough understanding of many laboratory calculations for they help in evaluating plant processes and efficiencies. Following are a few examples.

226. **If 50.4 grams (g) of calcium carbonate ($CaCO_3$) are dissolved in 1 liter of solution, how many moles of $CaCO_3$ were used, given that 1 mole equals 100.09 grams?**

Equation: $\mathbf{Moles = \dfrac{Grams\ of\ chemical}{Gram\ formula\ weight}}$

$Moles = \dfrac{50.4\ g}{100.09\ g/mole} = 0.5035$ moles, round to **0.504 moles of $CaCO_3$**

227. **Given the following data, what is the number of grams (g) in 1 mole of sulfuric acid (H_2SO_4)?**

Hydrogen (H) = 1.008 g/mole
Sulfur = 32.064 g/mole
Oxygen = 15.999 g/mole

Equation: **Grams/mole = 2(H g/mole) + S g/mole + 4(O g/mole)**

Grams/mole = 2(1.008 g/mole) + 32.064 g/mole + 4(15.999 g/mole)

Grams/mole = 2.016 g/mole + 32.064 g/mole + 63.996 g/mole = **98.076 g/mole for H_2SO_4**

228. Calculate the unseeded BOD$_5$ in mg/L, given the following data:

Start of test bottle dissolved oxygen (DO) = 9.0 mg/L
Bottle was incubated for 5 days in the dark at 20°C
After 5 days DO = 3.6 mg/L
Sample size = 100 mL
Total volume = 300 mL

Give answer to three significant figures.

Equation: **BOD$_5$ unseeded, mg/L =**

$$\frac{(\text{Initial DO, mg/L} - \text{Final DO, mg/L})(\text{Total volume, mL})}{\text{Sample volume, mL}}$$

$$\text{BOD}_5 \text{ unseeded, mg/L} = \frac{(9.0 \text{ mg/L} - 3.6 \text{ mg/L})(300 \text{ mL})}{100 \text{ mL}} = \textbf{16.2 mg/L unseeded BOD}_5$$

229. Calculate the seeded BOD$_5$ in mg/L, given the following data:

Sample size = 125 mL
Initial DO = 8.6 mg/L
Final DO = 3.2 mg/L
BOD$_5$ of seed stock = 95 mg/L
Seed stock = 5.0 mL
Total volume = 300 mL

First, calculate the seed correction in mg/L.

Equation: **Seed correction, mg/L =** $\dfrac{(\text{BOD}_5 \text{ of seed stock, mg/L})(\text{Seed stock, mg/L})}{\text{Total volume, mL}}$

$$\text{Seed correction, mg/L} = \frac{(95 \text{ mg/L})(5.0 \text{ mg/L})}{300 \text{ mL}} = 1.58 \text{ mg/L}$$

Next, calculate the BOD$_5$ seeded in mg/L.

Equation: **BOD$_5$ seeded, mg/L =**

$$\frac{(\text{Initial DO, mg/L} - \text{Final DO, mg/L} - \text{Seed correction, mg/L})(\text{Total volume, mL})}{\text{Sample volume, mL}}$$

$$\text{BOD}_5 \text{ seeded, mg/L} = \frac{(8.6 \text{ mg/L} - 3.2 \text{ mg/L} - 1.58 \text{ mg/L})\,(300 \text{ mL})}{125 \text{ mL}}$$

$$\text{BOD}_5 \text{ seeded, mg/} = \frac{(3.82 \text{ mg/L})\,(300 \text{ mL})}{125 \text{ mL}} = 9.168 \text{ mg/L, round to } \textbf{9.2 mg/L Seeded BOD}_5$$

BASIC ELECTRICITY PROBLEMS

Operators should have a basic understanding of electrical calculations, and they must always exercise safety in dealing with electricity at wastewater treatment plants or anywhere.

230. What is the voltage (E) on a circuit, if the current is 7 amperes (I) and the resistance (R) is 17 ohms? Give answer to three significant figures.

Equation: **Voltage = (Amps)(Resistance, ohms)**

Substitute values and solve.

Voltage = (7 amps)(17 ohms) = **119 volts**

231. What is the resistance on a circuit if the voltage is 120 and the amperes are 19?

Equation: **Resistance, ohms = Voltage/Amps**

Substitute values and solve.

Resistance, ohms = 120 Volts/19 Amps = **6.3 ohms**

1. If 98.5 lb of magnesium hydroxide [$Mg(OH)_2$] is dissolved in 248 gallons of water, what is the percent strength of the $Mg(OH)_2$ solution?

2. If the percent total solids are 5.38% and the dried sample (total solids) weighed 4.01 grams, what must have been the weight of the sludge sample before it was dried?

3. Calculate the volume in cubic feet for a pipeline that is 14.0 inches in diameter and 1,840 ft long.

4. A chemical tank is conical at the bottom and cylindrical at the top. If the diameter of the cylinder is 27.8 ft with a depth of 39.9 ft, and the cone depth is 12.6 ft, what is the volume of the tank in cubic feet and gallons?

5. Given the following parameters, calculate the solids loading rate on a secondary clarifier:

Clarifier diameter = 80.6 ft
Activated sludge flow = 2.36 mgd
Mixed liquor suspended solids (MLSS) = 3,560 mg/L

6. If the plant flow averages 1.76 mgd and production of biosolids averages 12,200 gal/day, what is the biosolids production in wet tons per year?

7. Determine the waste activated sludge (WAS) pumping rate in gpm given the following data:

Amount of WAS to be wasted = 4,225 lb/day
WAS suspended solids concentrations = 4,310 mg/L

8. How much sulfur dioxide in lb/day needs to be applied to dechlorinate a wastewater, if the flow is 3.09 mgd, the chlorine residual is 1.75 mg/L, and the sulfur dioxide must be 3.5 mg/L higher than the chlorine residual?

9. What is the number of lb/day of alum used by a wastewater plant given the following data?

Plant's treatment flow = 1,570 gpm
Alum dose = 9.38 mg/L
Alum = 48.5% aluminum sulfate

10. How many gpd of a 12.5% sodium hypochlorite solution are needed to disinfect a flow of 1,750,000 gallons, if the dosage required is 10.5 mg/L? Assume the solution weighs 8.34 lb/gal.

11. What should the chemical feeder be set on in mL/min, if the desired polymer dosage is 26.8 gpd?

12. Determine the feed rate of dry alum in lb/day, if the drawdown in exactly 5 minutes was 103.56 grams (g) and the flow is 1.20 mgd.

13. A wastewater plant with an influent flow of 2.37 mgd has primary influent suspended solids of 211 mg/L. If the secondary suspended solids are 156 mg/L, what is the amount of dry solids produced in lb/day?

14. A wastewater treatment pond receives a flow of 418,000 gpd. What is the organic loading rate in pounds of biochemical oxygen demand per day per acre (lb BOD_5/d/acre), if the pond has a surface area of 6.76 acre-ft and the influent BOD_5 concentration is 212 mg/L?

15. What is the soluble BOD_5 if the suspended solids are 139 mg/L, the total BOD_5 is 192 mg/L, and the K factor is 0.55?

16. Determine the amount of chemical oxygen demand (COD) entering an aeration tank in mg/L, if the influent flow is 1.84 mgd and the COD loading is 2,495 lb/day.

17. What is the hydraulic digestion time for a 58.7-ft diameter digester with a level of 19.4 ft and sludge flow of 20,850 gallons per day (gpd)?

18. What must have been the gas production by a digester in ft³/day given the following data?

Volatile solids destroyed = 750 lb/day
Gas produced in ft³/lb VS destroyed = 11.5 ft³/lb

19. Given the following data, calculate the mean cell residence time (MCRT) for this activated sludge system:

Aeration tank volume = 0.76 mil gal
Final clarifier volume = 0.20 mil gal
Mixed liquor suspended solids (MLSS) = 3,675 mg/L
Suspended solids (SS) wasted = 2,420 lb/day
Secondary effluent SS = 394 lb/day

20. What must have been the concentration of volatile acids in mg/L for a sour digester with a volume of 301,000 gallons, if the number of pounds of lime to neutralize the volatile acids was 4.580 lb?

21. Given the following data, determine the amount in lb/day removal of biochemical oxygen demand (BOD_5) from a trickling filter.

Plant influent flow = 2.05 mgd
Influent BOD_5 concentration = 313 mg/L
Effluent BOD_5 concentration = 174 mg/L

22. Given the following data on an aeration tank, calculate the current food-to-microorganism (F/M) ratio.

Primary effluent flow = 2.89 mgd
Volume of aeration tank = 239,000 gallons
MLVSS = 2,270 mg/L
BOD_5 = 203 mg/L

23. What is the solids loading for a dissolved air flotation (DAF) unit in lb/hr/ft² that is 62.8 ft by 19.9 ft, with a sludge flow of 0.225 mgd, a waste-activated sludge (WAS) concentration of 6,650 mg/L, and the sludge weighs 8.61 lb/gal?

24. If a DAF unit receives air at an average rate of 7.73 ft³/min, how many lb/hr of air does it receive?

25. A drying bed is 295 feet long and 38.8 feet wide. If 5.75 inches of sludge were applied to the drying bed, how many gallons of sludge were applied?

1.   **If 98.5 lb of magnesium hydroxide [Mg(OH)$_2$] is dissolved in 248 gallons of water, what is the percent strength of the Mg(OH)$_2$ solution?**

First, convert the number of gallons of water to pounds.

Number of lb = (248 gallons)(8.34 lb/gal) = 2,068.32 lb of water

Next, find the percent strength of the solution.

Equation: **Percent strength** $= \dfrac{(\text{Number of lb of chemical})(100\%)}{\text{Number of lb, Water} + \text{lb chemical}}$

Substitute values and solve.

Percent strength $= \dfrac{[98.5 \text{ lb Mg(OH)}_2](100\%)}{2,068.32 \text{ lb, Water} + 98.5 \text{ lb Mg(OH)}_2}$

Percent strength $= \dfrac{[98.5 \text{ lb Mg(OH)}_2](100\%)}{2,166.82 \text{ lb}} = $ **4.55% Mg(OH)$_2$ Solution**

2. **If the percent total solids are 5.38% and the dried sample (total solids) weighed 4.01 grams, what must have been the weight of the sludge sample before it was dried?**

Equation: **Percent total solids** $= \dfrac{(\text{Dry sample in grams})(100\%)}{\text{Sludge sample in grams}}$

Rearrange the equation.

Sludge sample, grams $= \dfrac{(\text{Dry sample in grams})(100\%)}{\text{Percent total solids}}$

Substitute values and solve.

Sludge sample, grams $= \dfrac{(4.01 \text{ grams})(100\%)}{5.38\%} = 74.54$ grams, round to **74.5 grams**

3. **Calculate the volume in cubic feet for a pipeline that is 14.0 inches in diameter and 1,840 ft long.**

First, convert the diameter to feet:

$(14.0 \text{ inches})\dfrac{(1 \text{ ft})}{12 \text{ in.}} = 1.1667$ ft (Diameter)

Then, convert the diameter to the radius:

Radius $=$ Diameter $/2 = 1.1667$ ft$/2 = 0.583$ ft (radius)

Formula for the volume of a pipe in cubic feet is:

πr^2**(Length)** or **(0.785)(Diameter)^2(Length)**:

Using the first equation, the Volume, ft^3 $= (3.14)(0.583 \text{ ft})(0.583 \text{ ft})(1,840 \text{ ft})$

Volume, ft^3 $= 1,963.74$ ft^3, round to **1,960 ft^3**

4. **A chemical tank is conical at the bottom and cylindrical at the top. If the diameter of the cylinder is 27.8 ft, with a depth of 39.9 ft, and the cone depth is 12.6 ft, what is the volume of the tank in cubic feet and gallons?**

First, find the volume of the cone in cubic feet.

Volume, $ft^3 = 1/3\pi r^2$(Depth):

Where the radius = Diameter/2 = 27.8 ft/2 = 13.9 ft

Volume, ft^3 = 1/3(3.14)(13.9 ft)(13.9 ft)(12.6 ft) = 2,548.05 ft^3

Next, find the volume of the cylindrical part of the tank.

Volume, $ft^3 = \pi r^2$(Depth) = (3.14)(13.9 ft)(13.9 ft)(39.9 ft) = 24,206.51 ft^3

Then, add the two volumes for the answer.

Total volume, ft^3 = 2,548.05 ft^3 + 24,206.51 ft^3 = 26,754.56 ft^3, round to **26,800 ft^3**

To find the number of gallons, multiply the total number of cubic feet by 7.48 gal/ft^3.

Number of gallons = (26,754.56 ft^3)(7.48 gal/ft^3) = 200,124.1 gal, round to **200,000 gal**

5. Given the following parameters, calculate the solids loading rate on a secondary clarifier:

Clarifier diameter = **80.6 ft**
Activated sludge flow = **2.36 mgd**
Mixed liquor suspended solids (MLSS) = **3,560 mg/L**

First, determine the area of the clarifier.

Area $= \pi r^2$ where r $=$ Diameter/2 $=$ 80.6 ft/2 $=$ 40.3 ft

Area $=$ (3.14)(40.3 ft)^2 = 5,099.64 ft^2

Next, calculate the solids loading rate.

Equation: **Solids loading rate** $= \dfrac{(\text{MLSS, mg/L})(\text{mgd})(8.34\ \text{lb/gal})}{\text{Area, ft}^2}$

Solids loading rate $= \dfrac{(3,560\ \text{mg/L})(2.36\ \text{mgd})(8.34\ \text{lb/gal})}{5,099.64\ \text{ft}^2}$

Solids loading rate $=$ 13.74 lb, round to **13.7 lb of solids/d/ft²**

6. If the plant flow averages 1.76 mgd and production of biosolids averages 12,200 gal/day, what is the biosolids production in wet tons per year?

Equation: **Biosolids, wet tons/yr** $= \dfrac{(\text{Biosolids, lb/mil gal})(\text{mgd})(365\ \text{days/yr})}{2,000\ \text{lb/ton}}$

Substitute values and solve.

Biosolids, wet tons/yr $= \dfrac{(12,200\ \text{lb/mil gal})(1.76\ \text{mgd})(365\ \text{days/yr})}{2,000\ \text{lb/ton}}$

Biosolids, wet tons/yr $=$ 3,918.64 wet tons/yr, round to **3,920 wet tons/yr**

7. Determine the waste activated sludge (WAS) pumping rate in gpm given the following data:

Amount of WAS to be wasted = 4,225 lb/day
WAS suspended solids concentrations = 4,310 mg/L

Use the "pounds" equation to solve this problem.

Equation: **Number of lb/day WAS = (WAS, mg/L)(Number of mgd)(8.34 lb/gal)**

Rearrange the equation to solve for mg/L.

$$\textbf{Number of mgd} = \frac{\text{Number of lb/day WAS}}{(\text{Number of mg/L WAS})(8.34 \text{ lb/gal})}$$

Substitute values and solve.

$$\text{Number of mgd} = \frac{4,225 \text{ lb/day}}{(4,310 \text{ mg/L WAS})(8.34 \text{ lb/gal})} = 0.1175 \text{ mgd}$$

Lastly, convert mgd to gpm.

$$\text{WAS pumping rate, gpm} = \frac{(0.1175 \text{ mgd})(1,000,000/\text{M})}{1,440 \text{ min/day}} = \textbf{81.6 gpm}$$

8. **How much sulfur dioxide in lb/day needs to be applied to dechlorinate a wastewater, if the flow is 3.09 mgd, the chlorine residual is 1.75 mg/L, and the sulfur dioxide must be 3.5 mg/L higher than the chlorine residual?**

First, determine how many mg/L of sulfur dioxide must be applied. This is the chlorine residual plus the amount that is required higher than the chlorine residual.

SO_2, mg/L = 1.75 mg/L + 3.5 mg/L = 5.25 mg/L SO_2

Next, determine the number of lb/day of SO_2 needed.

Equation: **Number of lb/day SO_2 = (SO_2, mg/L)(Number of mgd)(8.34 lb/gal)**

Substitute values and solve.

Number of lb/day SO_2 = (5.25 mg/L)(3.09 mgd)(8.34 lb/gal)

Number of lb/day SO_2 = 135.3 lb/day, round to **135 lb/day SO_2**

9. **What is the number of lb/day of alum used by a wastewater plant, given the following data?**

Plant's treatment flow = 1,570 gpm
Alum dose = 9.38 mg/L
Alum = 48.5% aluminum sulfate

First, convert gpm to mgd.

$$\text{Number of mgd} = \frac{(1,570 \text{ gpm})(1,440 \text{ min/day})}{1,000,000/\text{mil}} = 2.2608 \text{ mgd}$$

Next, calculate the number of lb/day of alum required.

Equation: $$\textbf{lb/day} = \frac{(\text{Alum, mg/L})(\text{mgd})(8.34 \text{ lb/gal})}{\text{Percent purity}/100\%}$$

$$\text{Alum, lb/day} = \frac{(9.38 \text{ mg/L})(2.2608 \text{ mgd})(8.34 \text{ lb/gal})}{48.5\%/100\%} = \textbf{365 lb/day of Alum}$$

10. **How many gpd of a 12.5% sodium hypochlorite solution are needed to disinfect a flow of 1,750,000 gallons, if the dosage required is 10.5 mg/L? Assume the solution weighs 8.34 lb/gal.**

First, convert gpd to mgd.

$$\text{Number of mgd} = \frac{1,750,000 \text{ gpd}}{1,000,000/\text{mil}} = 1.75 \text{ mgd}$$

Next, using the "pounds equation," calculate the lb day of chlorine needed.

Equation: **Chlorine, lb/day = (Dosage, mg/L)(mgd)(8.34 lb/gal)**

Chlorine, lb/day = (10.5 mg/L)(1.75 mgd)(8.34 lb/gal) = 153.25 lb/day

Because the solution is not 100%, divide the percent hypochlorite into the lb/day of chlorine needed.

$$\text{Hypochlorite, lb/day} = \frac{153.25 \text{ lb/day}}{12.5\%/100\%} = 1,226 \text{ lb/day hypochlorite}$$

Lastly, determine the gpd of hypochlorite solution needed.

$$\text{Hypochlorite, gpd} = \frac{1,226 \text{ lb/day}}{8.34 \text{ lb/gal}} = \textbf{147 gpd of Sodium hypochlorite}$$

11. **What should the chemical feeder be set on in mL/min, if the desired polymer dosage is 26.8 gpd?**

$$\text{Equation: } \textbf{Number of mL/min} = \frac{(\text{Number of gallons used}) (3,785 \text{ mL/gal})}{1,440 \text{ min/day}}$$

Substitute values and solve.

$$\text{Polymer, mL/min} = \frac{(26.8 \text{ gal}) (3,785 \text{ mL/gal})}{1,440 \text{ min/day}} = 70.44 \text{ mL/min, round to } \textbf{70.4 mL/min}$$

12. **Determine the feed rate of dry alum in lb/day, if the drawdown in exactly 5 minutes was 103.56 grams (g) and the flow is 1.20 mgd.**

First, determine the number of grams used per minute.

Alum, g = 103.56 g/5 min = 20.712 g/min

Know: 454 grams = 1 pound

Equation: **Alum, lb/day** $= \dfrac{(\text{Number of g/min})\,(1{,}440\ \text{min/day})}{454\ \text{g/lb}}$

Substitute values and solve.

Alum, lb/day $= \dfrac{(20.712\ \text{g/min})\,(1{,}440\ \text{min/day})}{454\ \text{g/lb}} = 65.69$ lb/day, round to **65.7 lb/day Alum**

13. **A wastewater plant with an influent flow of 2.37 mgd has primary influent suspended solids of 211 mg/L. If the secondary suspended solids are 156 mg/L, what is the amount of dry solids produced in lb/day?**

First, determine the number of mg/L of suspended solids (SS) removed.

SS removed, mg/L = 211 mg/L, influent − 156 mg/L effluent = 55 mg/L SS removed

SS removed, lb/day = (mg/L SS removed)(Number of mgd)(8.34 lb/gal)

SS removed, lb/day = (55 mg/L SS)(2.37 mgd)(8.34 lb/gal)

SS removed, lb/day = 1,087.12 lb/day, round to **1,090 lb/day of SS removed**

14. **A wastewater treatment pond receives a flow of 418,000 gpd. What is the organic loading rate in pounds of biochemical oxygen demand per day per acre (lb BOD$_5$/d/acre), if the pond has a surface area of 6.76 acre-ft and the influent BOD$_5$ concentration is 212 mg/L?**

First, convert gallons per day to mgd.

$$\text{Number of mgd} = \frac{418,000 \text{ gpd}}{1,000,000/\text{mil}} = 0.418 \text{ mgd}$$

Next, determine the pounds of BOD$_5$/d/acre using a modified version of the "pounds" equation.

Organic loading rate, lb BOD$_5$/d/acre $= \dfrac{(\text{BOD}_5, \text{mg/L}) (\text{Flow, mgd}) (8.34 \text{ lb/gal})}{\text{Surface area of pond, acre-ft}}$

Organic loading rate, lb BOD$_5$/d/acre $= \dfrac{(212 \text{ mg/L BOD}_5) (0.418 \text{ mgd}) (8.34 \text{ lb/gal})}{6.76 \text{ acre-ft}}$

Organic loading rate, lb BOD$_5$/d/acre = 109.33 lb BOD$_5$/d/acre, round to **109 lb BOD$_5$/d/acre**

15. **What is the soluble BOD$_5$ if the suspended solids are 139 mg/L, the total BOD$_5$ is 192 mg/L, and the K factor is 0.55?**

Equation: **Total BOD$_5$ = (Particulate BOD$_5$)(K factor) + Soluble BOD$_5$**

Rearrange the equation to solve for soluble BOD$_5$.

Soluble BOD$_5$ = Total BOD$_5$ − (Particulate BOD$_5$)(K factor)

Substitute values and solve.

Soluble BOD$_5$ = 192 mg/L BOD$_5$ − (139 mg/L BOD$_5$)(0.55 K factor)

Soluble BOD$_5$ = 192 mg/L BOD$_5$ − 76.45 mg/L BOD$_5$

Soluble BOD$_5$ = 115.55 mg/L, round to **120 mg/L Soluble BOD$_5$**

16. **Determine the amount of chemical oxygen demand (COD) entering an aeration tank in mg/L, if the influent flow is 1.84 mgd and the COD loading is 2,495 lb/day.**

Equation: **Number of lb/day COD = (COD, mg/L)(Number of mgd)(8.34 lb/gal)**

Rearrange the equation to solve for mg/L.

$$\text{Number of mg/L, COD} = \frac{\text{Number of lb/day COD}}{(\text{Number of mgd})(8.34\ \text{lb/gal})}$$

Substitute values and solve.

$$\text{Number of mg/L, COD} = \frac{2,495\ \text{lb/day}}{(1.84\ \text{mgd})(8.34\ \text{lb/gal})} = 162.59\ \text{mg/L, round to } \textbf{163 mg/L COD}$$

17. **What is the hydraulic digestion time for a 58.7-ft diameter digester with a level of 19.4 ft and sludge flow of 20,850 gallons per day (gpd)?**

First, determine the volume of the digester in ft³.

Volume, ft³ = (0.785)(Diameter)²(Depth, ft)

Volume, ft³ = (0.785)(58.7 ft)(58.7 ft)(19.4 ft) = 52,474.4 ft³

Next, determine the number of gallons in the digester.

Know: 1 ft³ = 7.48 gal/ft³

Number of gal = (52,474.4 ft³)(7.48 gal/ft³) = 392,508.5 gal

Lastly, calculate the digestion time in days.

Equation: **Digestion time, days** $= \dfrac{\text{Number of gallons}}{\text{Influent sludge flow, gal/day}}$

Substitute values and solve.

$$\text{Digestion time, days} = \frac{392,508.5\ \text{gal}}{20,850\ \text{gal/day}} = 18.825\ \text{days, round to } \textbf{18.8 days}$$

18. **What must have been the gas production by a digester in ft³/day given the following data?**

Volatile solids (VS) destroyed = 750 lb/day
Gas produced in ft³/lb VS destroyed = 11.5 ft³/lb

Equation: **Gas produced, ft³/lb VS destroyed** $= \dfrac{\text{Gas production, ft}^3/\text{day}}{\text{VS destroyed, lb/day}}$

Rearrange to solve for gas production in ft³/day.

Gas production, ft³/day = (Gas produced, ft³/lb VS destroyed)(VS destroyed, lb/day)

Substitute values and solve.

Gas production, ft³/day = (11.5 ft³/lb)(750 lb/day) = 8,625 ft³/day, round to **8,620 ft³/day**

19. **Given the following data, calculate the mean cell residence time (MCRT) for this activated sludge system:**

Aeration tank volume = 0.76 mil gal
Final clarifier volume = 0.20 mil gal
Mixed liquor suspended solids (MLSS) = 3,675 mg/L
Suspended solids (SS) wasted = 2,420 lb/day
Secondary effluent SS = 394 lb/day

First, add the aerator and final clarifier volumes.

Total volume, mil gal = 0.76 mil gal + 0.20 mil gal = 0.96 mil gal

Equation: **MCRT, days** $= \dfrac{(\text{MLSS, mg/L})(\text{mil gal})(8.34\ \text{lb/day})}{\text{SS wasted, lb/day} + \text{SS lb/day}}$

Substitute values and solve.

$$\text{MCRT, days} = \dfrac{(3,675\ \text{mg/L})(0.96\ \text{mil gal})(8.34\ \text{lb/day})}{2,420\ \text{lb/day SS} + 394\ \text{lb/day SS}}$$

$$\text{MCRT, days} = \dfrac{(3,675\ \text{mg/L})(0.96\ \text{mil gal})(8.34\ \text{lb/day})}{2,814\ \text{lb/day SS}} = 10.46\ \text{days, round to } \textbf{10 days}$$

20. **What must have been the concentration of volatile acids in mg/L for a sour digester with a volume of 301,000 gallons, if the number of pounds of lime to neutralize the volatile acids was 4.580 lb?**

Know: 1 mg/L of lime will neutralize 1 mg/L of volatile acids

First, convert the digester's volume from gallons to mil gal.

$$\text{Number of mil gal} = \frac{301,000 \text{ gal}}{1,000,000/\text{mil}} = 0.301 \text{ mil gal}$$

Equation: **Number of lb Volatile acids = (Volatile acids, mg/L)(mil gal)(8.34 lb/gal)**

Rearrange the equation to solve for the concentration of volatile acids.

$$\text{Volatile acids, mg/L} = \frac{\text{Number of lb, lime}}{(\text{mil gal})(8.34 \text{ lb/gal})}$$

Substitute values and solve.

$$\text{Volatile acids, mg/L} = \frac{4,580 \text{ lb, lime}}{(0.301 \text{ mil gal})(8.34 \text{ lb/gal})}$$

Volatile acids, mg/L = 1,824.45 mg/L, round to **1,820 mg/L Volatile acids**

21. **Given the following data, determine the amount in lb/day removal of biochemical oxygen demand (BOD$_5$) from a trickling filter.**

Plant influent flow = 2.05 mgd
Influent BOD$_5$ concentration = 313 mg/L
Effluent BOD$_5$ concentration = 174 mg/L

First, determine the amount of BOD$_5$ removed in mg/L by subtracting the influent BOD$_5$ from the effluent BOD$_5$.

BOD$_5$ removed, mg/L = (Influent BOD$_5$, mg/L − Effluent BOD$_5$, mg/L)

BOD$_5$ removed, mg/L = 313 mg/L − 174 mg/L = 139 mg/L BOD$_5$ removed

Next, solve the amount of BOD$_5$ removed in lb/day by using the "pounds" formula.

Equation: **Number of lb/day BOD$_5$ = (BOD$_5$, mg/L)(Number of mgd)(8.34 lb/gal)**

Substitute values and solve.

Number of lb/day BOD$_5$ removed = (139 mg/L BOD$_5$)(2.05 mgd)(8.34 lb/gal)

Number of lb/day BOD$_5$ removed = 2,376.48 lb/day, round to **2,380 lb/day BOD$_5$ removed**

22. **Given the following data on an aeration tank, calculate the current food-to-microorganism (F/M) ratio.**

Primary effluent flow = 2.89 mgd
Volume of aeration tank = 239,000 gallons
MLVSS = 2,270 mg/L
BOD$_5$ = 203 mg/L

First, convert the volume of wastewater in the tank to mil gal.

$$\text{Number of mil gal} = \frac{239,000 \text{ gal}}{1,000,000/\text{mil}} = 0.239 \text{ mil gal}$$

Next, write the equation: $$\textbf{F/M} = \frac{(\text{BOD}_5, \text{mg/L})(\text{Flow, mgd})(8.34 \text{ lb/gal})}{(\text{mg/L MLVSS})(\text{Volume in tank, mil gal})(8.34 \text{ lb/gal})}$$

The 8.34 lb/gal cancels, leaving the following equation:

$$\text{F/M} = \frac{(\text{BOD}_5, \text{mg/L})(\text{Flow, mgd})}{(\text{mg/L MLVSS})(\text{Volume of tank, mil gal})}$$

Substitute values and solve.

$$\text{F/M} = \frac{(203 \text{ mg/L BOD}_5)(2.89 \text{ mgd})}{(2,270 \text{ mg/L MLVSS})(0.239 \text{ mil gal})} = \textbf{1.08 F/M ratio}$$

23. **What is the solids loading for a dissolved air flotation (DAF) unit in lb/hr/ft² that is 62.8 ft by 19.9 ft, with a sludge flow of 0.225 mgd, a waste-activated sludge (WAS) concentration of 6,650 mg/L, and the sludge weighs 8.61 lb/gal?**

First, determine the area of the DAF unit in ft².

DAF area, ft² = (62.8 ft)(19.9 ft) = 1,249.72 ft²

Next, calculate the solids loading using the following equation.

Equation: **Solids loading, lb/hr/ft²** $= \dfrac{\text{(WAS, mg/L)(mgd)(lb/gal, Sludge)}}{\text{(DAF area, ft}^2\text{)(24 hr/day)}}$

Solids loading, lb/hr/ft² $= \dfrac{(6{,}650 \text{ mg/L, WAS})(0.225 \text{ mgd})(8.61 \text{ lb/gal, Sludge})}{(1{,}249.72 \text{ ft}^2 \text{ DAF})(24 \text{ hr/day})}$

Solids loading, lb/hr/ft² = 0.4295 lb/d/ft², round to **0.430 lb/hr/ft²**

24. **If a DAF unit receives air at an average rate of 7.73 ft³/min, how many lb/hr of air does it receive?**

Know: Air = 0.0807 lb/ft³ at standard temperature, pressure, and average composition

Equation: **Air, lb/day = (Number of ft³/min)(60 min/hr)(0.0807 lb/ft³)**

Substitute values and solve.

Air, lb/day = (7.73 ft³/min)(60 min/hr)(0.0807 lb/ft³)

Air, lb/day = 37.43 lb/hr, round to **37.4 lb/hr of air**

25. A drying bed is 295 ft long and 38.8 ft wide. If 5.75 inches of sludge were applied to the drying bed, how many gallons of sludge were applied?

First, convert 5.75 inches to feet.

Number of feet = 5.75 in./12 in./ft = 0.479 ft

Next, determine the volume in ft^3 sent to the drying bed.

Volume, ft^3 = (295 ft)(38.8 ft)(0.479 ft) = 5,482.63 ft^3

Lastly, calculate the volume in gallons sent to the sand drying beds.

Number of gal = (5,482.63 ft^3)(7.48 gal/ft^3) = 41,010 gal, round to **41,000 gal**

CHAPTER 2

WASTEWATER TREATMENT
Grade 2

Students preparing for the Grade 3 and Grade 4 wastewater treatment certification tests should also understand these problems.

COMMON MISCELLANEOUS CONVERSION PROBLEMS

These problems are a good refresher for the student or operator. They are essential to know for they are used constantly in wastewater calculations.

1. **Convert 20.8 million gallons per day (mgd) into cubic feet per second (ft³/s).**

Equation: $\textbf{Number of ft}^3\textbf{/s} = \dfrac{\text{(mgd)}\,(1,000,000\text{ gal})\,(1\text{ ft}^3)\,(1\text{ day})\,(1\text{ min})}{(1\text{ mil gal})\,(7.48\text{ gal})\,(1,440\text{ min})\,(60\text{ sec})}$

$\dfrac{(20.8\text{ mgd})\,(1,000,000\text{ gal})\,(1\text{ ft}^3)\,(1\text{ day})\,(1\text{ min})}{(1\text{ mil gal})\,(7.48\text{ gal})\,(1,440\text{ min})\,(60\text{ sec})} = \textbf{32.2 ft}^3\textbf{/s}$

2. **How many mil gal are there in 2,067 acre-ft?**

(2,067 acre-ft)(43,560 ft³/acre-ft)(7.48 gal/ft³)(1 mil/1,000,000) = **673 mil gal**

3.   Convert 52.4 ft³/s to mgd.

Equation: **Number of mgd** $= \dfrac{(\text{Number of ft}^3)(60 \text{ sec})(1{,}440 \text{ min})(7.48 \text{ gal})(1 \text{ mil gal})}{(\text{sec})(\text{min})(\text{day})(\text{ft}^3)(1{,}000{,}000 \text{ gal})}$

Number of mgd $= \dfrac{(52.4 \text{ ft}^3)(60 \text{ sec})(1{,}440 \text{ min})(7.48 \text{ gal})(1 \text{ mil gal})}{(\text{sec})(\text{min})(\text{day})(\text{ft}^3)(1{,}000{,}000 \text{ gal})} = \textbf{33.9 mgd}$

4.   How many gallons are there in 303 acre-ft?

First, convert acre-ft into cubic feet.

(303 acre-ft)(43,560 ft³/acre-ft) = 13,198,680 ft³

Substitute values and solve.

(13,198,680 ft³)(7.48 gal/ft³) = 98,726,126 gal, round to **98,700,000 gal**

5.   Convert 2,880 gpm to ft³/s.

Equation: **Number of ft³/s** $= \dfrac{\text{Number of gpm}}{(60 \text{ sec/min})(7.48 \text{ gal/ft}^3)}$

Number of ft³/s $= \dfrac{2{,}880 \text{ gpm}}{(60 \text{ sec/min})(7.48 \text{ gal/ft}^3)} = \textbf{6.42 ft}^3\textbf{/s}$

6.   Convert 7.55 ft³/s to gallons per day (gpd).

Equation: **Number of gpd = (Number of ft³/s)(86,400 s/day)(7.48 gal/ft³)**

Number of gpd = (7.55 ft³/s)(86,400 s/day)(7.48 gal/ft³)

Number of gpd = 4,879,354 gpd, round to **4,880,000 gpd**

7.   Convert 102 lb/mil gal to mg/L.

First, convert lb/mil gal concentration to lb, gal/mil gal, lb by dividing by 8.34 lb/gal.

$$\text{lb/mil lb} = \frac{(102\text{ lb})(1\text{ gal})}{(1\text{ mil gal})(8.34\text{ lb})} = \frac{102\text{ lb, gal}}{8.34\text{ lb, mil gal}}$$

The units of lb and gal cancel each other out leaving:

$$\frac{12.23}{\text{mil}} = 12.23\text{ mg/L, round to }\textbf{12.2 mg/L}$$

8.   Convert 10.8 ppm to lb/mil gal. Assume specific gravity equals 8.34 lb/gal.

10.8 ppm is the same as $\dfrac{10.8\text{ lb}}{1\text{ mil lb}}$

Now, convert lb/mil lb to lb/mil gal by multiplying by 8.34 lb/gal.

$$\frac{(10.8\text{ lb})(8.34\text{ lb/gal})}{1\text{ mil lb}} = 90.072\text{ lb/mil gal, round to }\textbf{90.1 lb/mil gal}$$

TEMPERATURE CONVERSION PROBLEMS

These problems are a good refresher for the student or operator. Temperature is important in many plant processes. In sedimentation, the higher the temperature of the water the faster the suspended particles, floc, and other fine materials will settle. Cold temperature water is denser than warmer water and thus particles take longer to settle. Microorganisms grow faster in warmer water than cold water, which affects processes, such as oxidation ditches, ponds, and digesters.

9.   Convert 12.4 degrees Celsius to degrees Fahrenheit.

Equation for Fahrenheit: **°F = 9°F/5°C(°C) + 32°F or °F = 1.8(°C) + 32°F**

°F = 9°F/5°C(12.4°C) + 32°F = 54.32°F, round to **54.3°F**

10. **Convert 72 degrees Fahrenheit to degrees Celsius.**

Equation for Celsius: **°C = 5°C/9°F(°F − 32°F)**

°C = 5°C/9°F(72°F − 32°F) = 5/9(40) = **22°C**

11. **Convert −12 degrees Celsius to degrees Fahrenheit.**

Equation: **°F = 9°F/5°C(°C) + 32°F**

°F = 9°F/5°C(−12 °C) + 32°F = −10.4°F, round to **−10.4°F**

12. **Convert 38 degrees Celsius to degrees Fahrenheit.**

Equation: **°F = 9°F/5°C(°C) + 32°F**

°F = 9°F/5°C(38°C) + 32°F = 100.4°F, round to **100°F**

RATIO CALCULATIONS

Ratios are a quick and easy way to solve simple problems when a particular relationship of two variables is known and one of those variables is changed to a known value. The question now is what happens to the variable that was not changed? The final result for this unknown variable can be calculated by setting up a ratio, that is, a "relationship" between the known variables can be set to equal the new "relationship" with the unknown. Then, using simple algebra, solve for the unknown. The following problems are examples of ratio problems that wastewater operators may find useful in their work.

13. **A wastewater treatment plant's digester unit produces 1,945 lb/day of solids at a flow of 15.4 gpm. If the flow increases to 18.5 gpm and all other parameters remain the same, what should be the number of lb/day of solids produced?**

Set up a ratio:

$$\frac{\text{Digester solids}_1, \text{lb/day}}{\text{Flow}_1, \text{gpm}} = \frac{\text{Digester solids}_2, \text{lb/day}}{\text{Flow}_2, \text{gpm}}$$

Rearrange the ratio to solve for digester solids$_2$.

$$\text{Digester solids}_2, \text{lb/day} = \frac{(\text{Digester solids}_1, \text{lb/day})\,(\text{Flow}_2, \text{gpm})}{\text{Flow}_1, \text{gpm}}$$

Substitute values and solve.

$$\text{Digester solids}_2, \text{lb/day} = \frac{(1,945 \text{ lb/day})\,(18.5 \text{ gpm})}{15.4 \text{ gpm}}$$

Digester solids$_2$, lb/day = 2,336.53 lb/day, round to **2,340 lb/day Solids**

14. **A chemical pump discharges 128 mL of alum at a speed setting of 48% and a stroke setting of 30%. If the alum pump's speed is increased to 62% and the stroke setting remains the same, what should be the mL output from the pump? Assume pump has a linear output.**

This problem can be solved using a ratio as follows:

$$\frac{\text{Alum dosage}_1, \text{mL}}{\text{Speed setting}_1, \%} = \frac{\text{Alum dosage}_2, \text{mL}}{\text{Speed setting}_2, \%}$$

Substitute values and solve.

$$\frac{128 \text{ mL}}{48\%} = \frac{x \text{ mL}}{62\%}$$

$$x \text{ mL} = \frac{(62\%)\,(128 \text{ mL})}{48\%} = 165.33 \text{ mL, round to } \textbf{170 mL Alum}$$

PERCENT AND PPM CONVERSION PROBLEMS

These problems are a good refresher for the student or operator. They are important because they are used in some dosage and mixture problems.

15. Convert a solution that has 305,000 ppm to percent.

A 1% solution = 10,000 ppm

$$\frac{305,000 \text{ ppm}}{10,000 \text{ ppm}/1\%} = \textbf{30.5\% Solution}$$

16. A solution was found to be 0.48% alum. What is the ppm alum in the solution?

If a 1% solution has 10,000 ppm then a 0.48% will have:

$(0.48\%)(10,000 \text{ ppm}/1\%) = \textbf{4,800 ppm Alum}$

17. Convert a solution that has 259,000 ppm to percent.

A 1% solution = 10,000 ppm

$$\frac{259,000 \text{ ppm}}{10,000 \text{ppm}/1\%} = \textbf{25.9\% Solution}$$

PERCENT CALCULATIONS

Percent calculations are used throughout this book and are thus essential to understand. They may also be a good refresher for the student or operator.

18. **What is the percent BOD_5 removal across a series of wastewater treatment ponds, if the influent BOD_5 to the first pond is 366 mg/L and the effluent BOD_5 from the last pond is 77 mg/L?**

Equation: **Percent BOD_5 removal** $= \dfrac{(\text{In} - \text{Out})(100\%)}{\text{In}}$

Percent BOD_5 removal $= \dfrac{(366\,\text{mg/L} - 77\,\text{mg/L})(100\%)}{366\,\text{mg/L}} = \dfrac{(289\,\text{mg/L})(100\%)}{366\,\text{mg/L}}$

Percent BOD_5 removal $= (0.7896)(100\%) = 78.96\,\%$, round to **79% BOD_5 removal**

19. **Ten pounds of lime are mixed in a 55-gallon drum that contains 50 gallons of water. What is the percent by weight of lime in the slurry?**

Percent lime $=$

$\dfrac{(10\,\text{lb})(100\%)}{10\,\text{lb} + (8.34\,\text{lb/gal})(50\,\text{gal})} = \dfrac{(10\,\text{lb})(100\%)}{10\,\text{lb} + 417\,\text{lb}} = \dfrac{(10\,\text{lb})(100\%)}{427\,\text{lb}} =$ **2% Slurry by weight**

20. **The biochemical oxygen demand (BOD_5) entering a trickling filter plant is 275 mg/L. What is the percent removal, if the final effluent water contains 27 mg/L BOD_5?**

Equation: **Percent BOD_5 removal** $= \dfrac{(\text{In} - \text{Out})(100\%)}{\text{In}}$

Percent BOD_5 removal $= \dfrac{(275\,\text{mg/L} - 27)(100\%)}{275\,\text{mg/L}} =$ **90% BOD_5 removal efficiency**

PERCENT STRENGTH BY WEIGHT SOLUTION PROBLEMS

The strength of solution calculations are important to determine so that operators can properly mix chemicals in the percentages they need for dosing a particular wastewater process or other application.

21. **If 50.0 lb of lime is dissolved in 100.0 gal of water, what is the percent strength by weight of the lime solution?**

First, convert the number of gallons of water to pounds.

Number of lb = (100.0 gal)(8.34 lb/gal) = 834 lb of water

Next, find the percent strength of the solution.

Equation: **Percent strength** $= \dfrac{(\text{Number of lb of chemical})(100\%)}{\text{Number of lb, Water} + \text{lb, Chemical}}$

Percent strength $= \dfrac{(50.0 \text{ lb Lime})(100\%)}{834 \text{ lb, Water} + 50.0 \text{ lb Lime}}$

Percent strength $= \dfrac{(50.0 \text{ lb Lime})(100\%)}{884 \text{ lb}} =$ **5.66% Lime solution by weight**

22. **If 501.2 grams (g) of magnesium hydroxide [Mg(OH)$_2$] are dissolved 10.0 liters (L) of water, what is the percent strength by weight of the solution?**

Equation: **Percent strength** $= \dfrac{(\text{Number of lb of chemical})(100\%)}{\text{Number of lb, Water} + \text{lb Chemical}}$

Substitute values and solve.

Percent strength $= \dfrac{[501.2 \text{ g Mg(OH)}_2](100\%)}{(10.0 \text{ L})(1{,}000 \text{ g/L}) + 501.2 \text{ g Mg(OH)}_2}$

Percent strength $= \dfrac{[501.2 \text{ g Mg(OH)}_2](100\%)}{10{,}501.2 \text{ g}} =$ **4.77% Mg(OH)$_2$ solution by weight**

23. If 501.2 lb of magnesium hydroxide are dissolved 10.0 gallons of water, what is the percent strength by weight of the solution?

Equation: **Percent strength** $= \dfrac{(\text{Number of lb, chemical})(100\%)}{(\text{Number of gal})(8.34\ \text{lb/gal}) + \text{Number of lb, Chemical}}$

Substitute values and solve.

Percent strength $= \dfrac{(501.2\ \text{lb})(100\%)}{(10.0\ \text{gal})(8.34\ \text{lb/gal}) + 501.2\ \text{lb}} = \dfrac{50,120\ \text{lb}\,\%}{83.4\ \text{lb} + 501.2\ \text{lb}}$

Percent strength $= \dfrac{50,120\ \text{lb}\,\%}{584.6\ \text{lb}} = 85.73\%$, round to **85.7% Mg(OH)$_2$ solution by weight**

PERCENT SOLIDS BY WEIGHT CALCULATIONS

Operators use percent solids calculations to determine efficiency of different unit processes, as well as determining how much waste will require disposal.

24. What is the percent by weight of total inorganic solids in a sludge sample given the following data?

Sludge sample wet weight = 354 g
Total solids dry weight = 27.3 g
Inorganic dry weight = 3.62 g

Equation: **Percent inorganic solids** $= \dfrac{(\text{Dry sample in grams})(100\%)}{\text{Sludge sample in grams}}$

Percent total solids $= \dfrac{(3.62\ \text{g})(100\%)}{354\ \text{g}} = $ **1.02% Total inorganic solids by weight**

25. **A wastewater treatment plant pumps an average of 30,250 gal/day of sludge. If the percent by weight of solids in the sludge averages 2.17%, what is the lb/day of solids pumped?**

First, calculate the number of lb/day of sludge pumped.

Sludge, lb/day = (30,250 gal/day)(8.34 lb/gal) = 252,285 lb/day

Next, calculate the lb/day of solids pumped.

Equation: **Solids, lb/day** $= \dfrac{(\text{Sludge, lb/day})(\text{Percent solids})}{100\%}$

Solids, lb/day $= \dfrac{(252,285 \text{ lb/day})(2.17\%)}{100\%} = 5,474.58$ lb/day, round to **5,470 lb/day**

PERCENT VOLATILE SOLIDS REDUCTION

The percent volatile solids reduction calculations indicate the effectiveness of the digested sludge process when compared to the volatile solids in the influent. The higher the percent volatile solids reduced or destroyed, the more stable the organic matter in the digester becomes and the more gas that is produced.

26. **If the sludge entering a digester has a volatile solids (VS) content of 58.1% by weight and the digester effluent sludge has a VS content of 41.5% by weight, calculate the percent VS reduction.**

First, convert percentage to decimal form by dividing by 100%.

58.1%/100% = 0.581 and 41.5%/100% = 0.415

Equation: **Percent VS reduction** =

$$\dfrac{(\text{Percent influent VS} - \text{Percent effluent VS})(100\%)}{[\text{Percent influent VS} - (\text{Percent influent VS})(\text{Percent effluent VS})]}$$

Percent VS reduction $= \dfrac{(0.581 - 0.415)(100\%)}{0.581 - (0.581)(0.415)} = \dfrac{0.166(100\%)}{0.581 - 0.241115}$

Percent VS reduction $= \dfrac{16.6\%}{0.339885} = 48.84\%$, round to **48.8% VS reduction by weight**

27. **Calculate the percent volatile solids (VS) reduction, if the digester influent sludge has a VS content of 60.1% by weight and the digester effluent sludge has a VS content of 39.9% by weight.**

First, convert percentage to decimal form by dividing by 100%.

60.1%/100% = 0.601 and 39.9%/100% = 0.399

Equation condensed: **Percent VS reduction** $= \dfrac{(\text{Influent} - \text{Effluent})(100\%)}{\text{Effluent} - (\text{Effluent})(\text{Influent})}$

Percent VS reduction $= \dfrac{(0.601 - 0.399)(100\%)}{0.601 - (0.601)(0.399)} = \dfrac{0.202(100\%)}{0.601 - 0.239799}$

Percent VS reduction $= \dfrac{20.2\%}{0.361201} = 55.92\%$, round to **55.9% VS reduction by weight**

28. **What is the percent volatile matter (VM) reduction for a digester, if the raw biosolids VM is 68.9% by weight and the VM digested biosolids is 49.6% by weight?**

Percent VM reduction =

$$\dfrac{(\text{Percent influent VM} - \text{Percent effluent VM})(100\%)}{[\text{Percent influent VM} - (\text{Percent influent VM})(\text{Percent effluent VM})]}$$

First, convert percentages to decimal form for easier substitution.

Raw biosolids = 68.9%/100% = 0.689

Digested VM = 49.6%/100% = 0.496

Substitute values and solve.

Percent reduction $= \dfrac{(0.689 - 0.496)(100\%)}{[0.689 - (0.689)(0.496)]}$

Simplify:

Percent reduction $= \dfrac{(0.193)(100\%)}{(0.689 - 0.341744)}$

Percent reduction $= \dfrac{(0.193)(100\%)}{0.347256} =$ **55.6% VM reduction by weight**

PERCENT MOISTURE REDUCTION PROBLEMS

This calculation will tell the operator the efficiency of the moisture reduction process.

29. **What is the percent moisture reduction for a digester, if the raw biosolids is 7.4% solids by weight and the digested biosolids solids is 14.2% by weight?**

Equation: **Percent moisture reduction =**

$$\frac{(\text{Percent influent moisture} - \text{Percent moisture, after digestion})(100\%)}{[\text{Percent influent moisture} - (\text{Percent influent moisture})(\text{Percent moisture, after digestion})]}$$

First, convert the percentages for solids to moisture percent then to decimal form for easier substitution.

Raw biosolids = 100% − 7.4% = 92.6%/100% = 0.926

Digested biosolids = 100% − 14.2% = 85.8%/100% = 0.858

Substitute values and solve.

$$\text{Percent moisture reduction} = \frac{(0.926 - 0.858)(100\%)}{[0.926 - (0.926)(0.858)]}$$

Simplify:

$$\text{Percent moisture reduction} = \frac{(0.068)(100\%)}{(0.926 - 0.794508)}$$

$$\text{Percent moisture reduction} = \frac{(0.068)(100\%)}{0.131492} = \textbf{52\% Moisture reduction by weight}$$

30. **What is the percent moisture reduction for a digester, if the raw biosolids is 8.6% solids by weight and the digested biosolids solids is 16.5% by weight?**

Equation: **Percent moisture reduction =**

$$\frac{(\text{Percent influent moisture} - \text{Percent moisture, after digestion})(100\%)}{[\text{Percent influent moisture} - (\text{Percent influent moisture})(\text{Percent moisture, after digestion})]}$$

First, convert the percentages for solids to moisture percent then to decimal form for easier substitution.

Raw biosolids $= 100\% - 8.6\% = 91.4\%/100\% = 0.914$

Digested biosolids $= 100\% - 16.5\% = 83.5\%/100\% = 0.835$

Substitute values and solve.

Percent moisture reduction $= \dfrac{(0.914 - 0.835)(100\%)}{[0.914 - (0.914)(0.835)]}$

Simplify:

Percent moisture reduction $= \dfrac{(0.079)(100\%)}{(0.914 - 0.76319)}$

Percent moisture reduction $= \dfrac{(0.079)(100\%)}{0.15081} = $ **52% Moisture reduction by weight**

VOLATILE SOLIDS PUMPING CALCULATIONS

These calculations are used as a planning tool by the operator. By knowing the pumping rate of volatile solids into a digester, an operator can make sure it is not overloaded, which would adversely affect the digester's operation and performance. See Figures 2, 4, 5, and 6 in Appendix E for four types of wastewater plants using a digester.

31. **Given the following data, how many lb/day of volatile solids (VS) are pumped to a digester?**

Pumping rate $= 4,890$ gpd
Solids content $= 5.83\%$
Volatile solids $= 62.1\%$
Specific gravity of sludge $= 1.07$

First, determine the lb/gal for the sludge.

Sludge, lb/gal $= (8.34 \text{ lb/gal})(1.07) = 8.92 \text{ lb/gal}$

Equation: **VS, lb/day** $= \dfrac{(\text{Number of gpd to digester})(\text{Percent solids})(\text{Percent VS})(8.34 \text{ lb/gal})}{(100\%)(100\%)}$

VS, lb/day $= \dfrac{(4,890 \text{ gpd Solids})(5.83\%)(62.1\% \text{ VS})(8.92 \text{ lb/gal})}{(100\%)(100\%)}$

VS, lb/day $= 1,579.19$ lb/day, round to **1,580 lb/day VS**

32. **Given the following data, how many lb/day of volatile solids (VS) are pumped to a digester?**

Pumping rate = 5.1 gpm
Solids content = 4.12%
Volatile solids = 57.5%
Specific gravity of sludge = 1.06

First, determine the lb/gal for the sludge.

Sludge, lb/gal = (8.34 lb/gal)(1.06) = 8.84 lb/gal

Next, convert gpm to gpd.

Number of gpd = (5.1 gpm)(1,440 min/day) = 7,344 gpd

Equation: **VS, lb/day =**

$$\frac{(\text{Number of gpd to digester}) \, (\text{Percent solids}) \, (\text{Percent VS}) \, (8.34 \, \text{lb/gal})}{(100\%) \, (100\%)}$$

$$\text{VS, lb/day} = \frac{(7,344 \, \text{gpd Solids}) \, (4.12\%) \, (57.5\% \, \text{VS}) \, (8.84 \, \text{lb/gal})}{(100\%) \, (100\%)}$$

VS, lb/day = 1,537.98 lb/day, round to **1,500 lb/day VS**

SOLUTION MIXTURE CALCULATIONS

These calculations are used when mixing two of the same solutions that have different strengths given a volume target. They are important for the operator to understand because there most probably will be times when solutions will require mixing.

33. **How many gallons of a 12.5% solution must be mixed with a 4.8% solution to make exactly 500 gallons of a 10.0% solution? Give answers to nearest gallon.**

There are two ways to solve dilution problems. The dilution triangle is perhaps the easiest, and is shown below for the next two problems.

How to solve the problem using the dilution triangle: The two numbers on the left are the existing concentrations of 12.5% and 4.8%. The number in the center, 10.0%, is the desired concentration. The numbers on the right are determined by subtracting diagonally the existing concentrations from the desired concentration.

12.5%		5.2*[1]	5.2 parts of the 12.5% solution are required for every 7.7 parts.
	10.0%		
4.8%		2.5*[2]	2.5 parts of the 4.8% solution are required for every 7.7 parts.
		7.7	total parts

*[1] 5.2 is determined by subtracting diagonally 4.8% from 10.0%.
*[2] 2.5 is determined by subtracting diagonally 12.5% from 10.0%. The negative sign is dropped.

$$\frac{5.2 \, \text{parts} \, (500 \, \text{gal})}{7.7 \, \text{parts}} = 337.66 \text{ gallons, round to } \textbf{338 gallons of the 12.5\% solution.}$$

$$\frac{2.5 \, \text{parts} \, (500 \, \text{gal})}{7.7 \, \text{parts}} = 162.34 \text{ gallons, round to } \underline{\textbf{162 gallons of the 4.8\% solution.}}$$

500 gallons added to cross check math.

To make the 500 gallons of the 10.0% solution, mix 338 gallons of the 12.5% solution with 162 gallons of the 4.8% solution.

34. **A solution containing 325 gallons of 6.5% hypochlorite is required. How many gallons of a 9.8% solution must be mixed with a 2.5% solution to make the required solution? Assume three significant figures for gallons to be mixed.**

Solve the problem using the dilution triangle.

9.8%		4.0	4.0 parts of the 9.8% solution are required for every 7.3 parts.
	6.5%		
2.5%		3.3	3.3 parts of the 2.5% solution are required for every 7.3 parts.
		7.3	total parts

$$\frac{4.0 \, \text{parts} \, (325 \, \text{gal})}{7.3 \, \text{parts}} = \textbf{178 gallons of the 9.8\% solution}$$

$$\frac{3.3 \, \text{parts} \, (325 \, \text{gal})}{7.3 \, \text{parts}} = \underline{\textbf{147 gallons of the 2.5\% solution}}$$

325 gallons

To make the 325 gallons of the 6.5% solution, mix 178 gallons of the 9.8% solution with 147 gallons of the 2.5% solution.

CALCULATIONS FOR ARITHMETIC MEAN, MEDIAN, RANGE, MODE, AND GEOMETRIC MEAN

These calculations are good tools for planning and evaluating plant processes.

35. What is the average number of pounds of lime used per day given the following data?

Mon.	Tues.	Wed.	Thurs.	Fri.	Sat.	Sun.
254	241	261	250	236	240	262

Equation: **Average lime used, lb/day** $= \dfrac{\text{Sum of lime used each day, lb}}{\text{Total time, days}}$

Substituting:

Avg. lime used, lb/day $= \dfrac{254 + 241 + 261 + 250 + 236 + 240 + 262}{7 \text{ days}} =$ **249 lb/day Lime**

36. What is the average mgd production for a treatment plant given the following data?

Mon.	Tues.	Wed.	Thurs.	Fri.	Sat.	Sun.
3.8	3.5	3.1	3.0	3.2	3.2	3.5

Equation: **Average mgd produced** $= \dfrac{\text{Sum of mgd used each day}}{\text{Total time, days}}$

Avg. mgd produced $= \dfrac{3.8 + 3.5 + 3.1 + 3.0 + 3.2 + 3.2 + 3.5}{7 \text{ days}} = 3.33$ mgd, round to **3.3 mgd**

37. **Calculate the moving (running) average for influent flow to a wastewater plant during days 12, 13, and 14 given the following data:**

1—1.09 mgd	9—0.998 mgd
2—1.22 mgd	10—0.999 mgd
3—1.02 mgd	11—1.09 mgd
4—0.992 mgd	12—1.09 mgd
5—0.988 mgd	13—1.16 mgd
6—0.982 mgd	14—1.13 mgd
7—0.976 mgd	15—1.04 mgd
8—0.983 mgd	16—0.993 mgd

Day 12 7-day moving average:

Add day 12 mgd to the previous 6 days.

$$\text{7-day average} = \frac{0.982 + 0.976 + 0.983 + 0.998 + 0.999 + 1.09 + 1.09}{7 \text{ days}} = \textbf{1.02 mgd}$$

Day 13 7-day moving average:

Add day 13 mgd to the previous 6 days.

$$\text{7-day average} = \frac{0.976 + 0.983 + 0.998 + 0.999 + 1.09 + 1.09 + 1.16}{7 \text{ days}} = \textbf{1.04 mgd}$$

Day 14 7-day moving average:

Add day 14 mgd to the previous 6 days.

$$\text{7-day average} = \frac{0.983 + 0.998 + 0.999 + 1.09 + 1.09 + 1.16 + 1.13}{7 \text{ days}} = \textbf{1.06 mgd}$$

38. Given the following data, calculate the unknowns.

Note: A scientific calculator is required for determining the geometric mean.

Day	Effluent BOD_5, mg/L	Unknown
Monday	25	a. Arithmetic mean, mg/L
Tuesday	27	b. Median, mg/L
Wednesday	21	c. Range, mg/L
Thursday	20	d. Mode, mg/L
Friday	19	e. Geometric Mean, mg/L
Saturday	19	
Sunday	17	

Assume all measurements are to two significant figures.

a. Calculate the arithmetic mean of BOD_5 in mg/L to three significant figures.

Equation: **Arithmetic mean** $= \dfrac{\text{Sum of all measurements}}{\text{Number of measurements}}$

Arithmetic Mean $= \dfrac{25 + 27 + 21 + 20 + 19 + 19 + 17}{7} =$ **21.1 mg/L BOD_5**

b. Determine the median of BOD_5 mg/L.

To determine the median, put the chlorine dosages in ascending order and choose the middle value.

1	2	3	**4**	5	6	7
17	19	19	**20**	21	25	27

In this case, the middle value is **20 mg/L BOD_5.**

c. Determine the range of BOD_5 mg/L.

Equation: **Range = Largest value − Smallest value**

Range, mg/L = 27 mg/L − 17 mg/L = **10 mg/L BOD_5**

d. Determine the mode of BOD$_5$ mg/L.

Mode is the measurement that occurs most frequently.

In this case, it is **19 mg/L BOD$_5$.**

e. Calculate the geometric mean of BOD$_5$ to three significant figures.

Equation: **Geometric mean $= [(x_1)(x_2)(x_3)(x_4).....(x_n)]^{1/n}$**

Geometric mean, mg/L $= [(17)(19)(19)(20)(21)(25)(27)]^{1/7}$

Geometric mean, mg/L $= (1739839500)^{1/7}$

Geometric mean, mg/L $=$ **20.9 mg/L BOD$_5$**

AREA PROBLEMS

Areas are important to determine for a number of reasons including knowing the "footprint" of a tank or pond or the area of a particular process to make further calculations in other wastewater problems.

39. What is the area of a drying bed that is 250 ft long and 45 ft wide?

Equation: **Area $=$ (Length)(Width)**

Area $=$ (250 ft)(45 ft) $=$ 11,250 ft^2, round to **11,000 ft^2**

40. What is the area a tank occupies, if it has a diameter of 48.0 ft?

Equation: **Area = πr^2**

Where r is the radius *or* **Area = (0.785)(Diameter)^2**

Note: Both equations will be used throughout this book.

Area of tank = (0.785)(48.0 ft)(48.0 ft) = 1,809 ft^2, round to **1,810 ft^2**

41. If the surface area of a rectangular drying bed is 22,500 sq ft and one side is 50 ft, what is the length of the other side? Assume 50 ft is measured to the nearest foot.

First, write the equation for finding the area of a rectangle.

Area = (Length)(Width)

Since the area is 22,500 ft and one side is 50 ft, assume that the 50 ft is the width (remember, by convention the width is the shorter of the two measurements). Then, solve for the length by rearranging the equation as follows:

$$\text{Length, ft} = \frac{\text{Area of } 22,500 \text{ ft}^2}{(50 \text{ ft})} = \textbf{450 ft}$$

42. What is the diameter of a tank, if the surface area is 4,542 ft^2?

Write the equation for finding the area of a circular tank.

Area = (0.785)(Diameter)^2 or A = $r^2\pi$ or A = D$^2\pi$/4

Solve for the diameter by rearranging the equation as shown:

$$D^2 = \frac{\text{Area}}{0.785} = \frac{4,542 \text{ ft}^2}{0.785} = 5,786 \text{ ft}^2$$

Then, take the square root of 5,786 ft^2

Then: Diameter = **76.1 ft**

43. **What is the surface area of a pond that averages 302 ft in length and 179 ft wide?**

Area, ft^2 = (Length)(Width) = (302 ft)(179 ft) = 54,058 ft^2, round to **54,100 ft^2**

CIRCUMFERENCE PROBLEMS

The circumference is important to know for calculating the area of a circular tank or the area of a particular process to make further calculations in a problem. For example, an operator needs to calculate the weir overflow rate on a circular clarifier: given the diameter, the circumference or length of the weir can be calculated, and thus the weir overflow rate can be determined from this.

44. **What is the circumference in feet of a tank that is 48 ft in diameter?**

Equation: **Circumference = π(Diameter)**

Circumference = (3.14)(48 ft) = 150.72 ft, round to **150 ft**

45. **If the circumference of a tank is 210 ft, what is the radius?**

Circumference = π(Diameter) or Circumference = 2rπ

210 ft = (3.14)(Diameter) or Diameter = 210 ft./3.14 = 66.9 ft

The radius is equal to the Diameter/2 or radius = 66.9/2 = 33.45 ft, round to **33 ft**

VOLUME PROBLEMS

Volumes are very important to determine because many problems in the wastewater field require the volume to be known before the rest of the calculations can be made. Knowing the volume of a particular process can also help the operator plan and make proper decisions in the treatment of wastewater.

46. A circular tank has a radius of 24 ft and is 30.0 ft high. What is the capacity of the tank in cubic feet?

The volume equation for a circular tank is **Volume = πr²(Height)**, where r is the radius.

Volume of tank in ft³ = (3.14)(24 ft)(24 ft)(30.0 ft) = 54,259.2 ft³, round to **54,000 ft³**

47. What is the volume in cubic feet of two basins with the following dimensions: 200 ft by 30 ft by 6 ft and 180 ft by 26 ft by 6 ft?

Always write the equation: **Volume = (Length)(Width)(Depth)** or **V = (L)(W)(D).**

Solve each basin separately and then add their results.

Basin 1, ft³ = (200 ft)(30 ft)(6 ft) = 36,000 ft³

Basin 2, ft³ = (180 ft)(26 ft)(6 ft) = 28,080 ft³

36,000 ft³ + 28,080 ft³ = 64080 ft³, round to **60,000 ft³**

48. A soda ash tank is conical at the bottom and cylindrical at the top. If the diameter of the cylinder is 18 ft with a depth of 32 ft and the cone depth is 15 ft, what is the volume of the tank in ft³?

First, find the volume of the cone in ft³.

Volume, ft³ = 1/3(0.785)(Diameter)²(Depth)

Volume, ft³ = 1/3(0.785)(18 ft)(18 ft)(15 ft) = 1,272 ft³

Next, find the volume of the cylindrical part of the tank.

Volume, ft³ = (0.785)(Diameter)²(Depth) = (0.785)(18 ft)(18 ft)(32 ft) = 8,139 ft³

Lastly, add the two volumes for the answer.

Total volume, ft³ = 1,272 ft³ + 8,139 ft³ = 9,411 ft³, round to **9,400 ft³**

49. **Determine the volume in cubic feet for a pipe with a diameter of 3.0 ft and a length of 14.3 miles.**

First, determine the number of feet in 14.3 miles.

(5,280 ft/mile)(14.3 miles) = 75,504 ft

Equation: **Volume, ft³ = (0.785)(Diameter)²(Length)**

Volume, ft³ = (0.785)(3.0 ft)(3.0 ft)(75,504 ft) = 533,436 ft³, round to **530,000 ft³**

50. **What is the volume of a conical tank in cubic feet that is 12 ft in diameter and 11 ft in height?**

Equation for a cone is **Volume = 1/3πr²(Height or Depth)**

First, find the radius: radius = Diameter/2 *or* D/2: radius = 12 ft/2 = 6 ft.

Using equation, substitute the known quantities.

Volume, ft³ = 1/3(3.14)(6 ft)(6 ft)(11 ft) = 414.48 ft³, round to **410 ft³**

Note: Because measurements are to two significant figures, then the answer must also be two significant figures.

51. **A lime tank is conical at the bottom and cylindrical at the top. If the diameter of the cylinder is 14 ft, has a depth of 24 ft, and the cone depth is 12 ft, what is the volume of the tank in ft³?**

First, find the volume of the cone in cubic feet.

Volume, ft³ = 1/3πr²(Depth)

Where the radius = Diameter/2 = 14 ft/2 = 7 ft

Volume, ft³ = 1/3(3.14)(7 ft)(7 ft)(12 ft) = 615 ft³

Next, find the volume of the cylindrical part of the tank.

Volume, ft³ = πr²(Depth) = (3.14)(7 ft)(7 ft)(24 ft) = 3,693 ft³

Lastly, add the two volumes for the answer.

Total volume, ft³ = 615 ft³ + 3,693 ft³ = 4,308 ft³, round to **4,300 ft³**

52. **Find the volume in gallons for two wastewater ponds that are 302 ft by 148 ft and two smaller waste ponds that are 198 ft by 95 ft. The depth of water in the two largest ponds averages 6.85 ft and for the two smaller ponds 5.75 ft.**

First, find the volume in gallons for the 2 largest ponds.

Volume, gal = (Length)(Width)(Depth)(7.48 gal/ft³)(Number of waste ponds)

Volume, gal = (302 ft)(148 ft)(6.85 ft)(7.48 gal/ft³)(2 ponds) = 4,580,267 gal

Next, find the volume in gallons for the 2 smallest ponds.

Then, add both volumes.

Volume, gal = (198 ft)(95 ft)(5.75 ft)(7.48 gal/ft³)(2 ponds) = 1,618,036 gal

Total Volume, gal = 6,198,303 gal

Round the above total volume to **6,200,000 gal**

DENSITY CALCULATIONS

The density of a substance is the amount of mass for a given volume. It is usually expressed as lb/gal or lb/ft^3 in the English system or as g/cm^3, kg/L, or kg/dm^3 in the metric system. Mass is defined as the quantity of matter as determined from Newton's second law of motion or by its weight. Weight is defined as the force that gravitation exerts upon a body and is equal to the mass of the body times the local acceleration of gravity.

53. What is the density in lb/gal of a 2.000-liter solution that weighs 1.45 lb?

First, convert mL to gallons: (2.000 liters)(1 gal/3.785 liters) = 0.5284.

Equation: **Density = Mass/Volume**

Density of solution = 1.45 lb/0.5284 gal = **2.74 lb/gal**

54. The density of an unknown substance is 2.96 grams/cm^3. How much space would this substance occupy in cm^3, if it weighed 3.88 lb?

First, convert the number of lb to grams (g).

Number of g = (Number of lb)(454 g/1 lb)

Substitution: Number of g = (3.88 lb)(454 g/1 lb) = 1,761.52 g

We know that 2.96 grams of the substance occupies 1 cm^3 by knowing its density. To get the space 1,761.52 grams occupies we only need to divide by the density.

Space occupied by substance $= \dfrac{1{,}761.52\text{ g}}{2.96\text{ g/cm}^3} =$ **595 cm^3**

55. **Find the number of gal/ft³ of a solution, if it weighs 99.2 lb/ft³.**

Divide the weight of the substance by 8.34 lb/gal to compare the substance to water.

Density $= \dfrac{99.2 \text{ lb/ft}^3}{8.34 \text{ lb/gal}} = 11.89$ gal/ft³, round to **11.9 gal/ft³**

56. **A substance weighs 1,825 grams (g) and occupies a space of 705 cubic centimeters (cm³). What is its density in g/cm³?**

Equation: **Density = Mass/Volume**

Density = 1,825 g/705 cm³ = **2.59 g/cm³**

57. **The specific gravity of a solution is 1.33. How many lb will 1 ft³ weigh?**

Rearranging the above equation: **lb/ft³ = (sp gr)(62.4 lb/ft³)**

Solution, lb/gal = (1.33)(62.4 lb/ft³) = **83.0 lb/ft³ Solution**

SPECIFIC GRAVITY OF LIQUIDS

Specific gravity compares the density of one substance to another. Water is the standard for liquids and is equal to 1.

58. **The density of an unknown liquid is 65.3 lb/ft³. What is its specific gravity (sp gr)?**

Know that water has a density of 62.4 lb/ft³. Divide the density of the unknown by the density of water.

Equation: **Specific gravity (sp gr) = Density of substance/Density of water**

Sp gr of Unknown Substance $= \dfrac{65.3 \text{ lb/ft}^3}{62.4 \text{ lb/ft}^3} = $ **1.05 sp gr**

59. What is the specific gravity (sp gr) for a solution that weighs 12.04 lb/gal?

Know that the density of water can also be expressed as lb/gal, or 8.34 lb/gal.

$$\text{Sp gr} = \frac{12.04 \text{ lb/gal}}{8.34 \text{ lb/gal}} = \textbf{1.44 sp gr}$$

SPECIFIC GRAVITY OF SOLIDS

Specific gravity compares the density of one substance to another. Water is the standard for solids and is equal to 1.

60. A piece of metal that weighs 7.6125 grams (g) in air is weighed in water and found to be 7.2181 grams. What is the specific gravity (sp gr) of this metal?

First, subtract the weight in air from the weight in water to determine the loss of weight in water.

Number of grams = 7.6125 g − 7.2181 g = 0.3944 g is weight loss in water

Next, find the specific gravity by dividing the weight of the metal in air by the weight loss in water.

Sp gr = 7.6125 g/0.3944 g = **19.301 sp gr**

What is this metal?

61. What is the specific gravity (sp gr) of a metal if it weighs 354 grams in air and weighs 265 grams in water?

First, subtract the weight in air from the weight in water to determine the loss of weight in water.

Number of kilograms = 354 g − 265 g = 89 g is weight loss in water

Next, find the specific gravity by dividing the weight of the metal in air by the weight loss in water.

Sp gr = 354 g/89 g = 3.9775, round to **3.98 sp gr**

PRESSURE PROBLEMS

Pressure is the measure of force against a surface and is usually expressed as force per unit area. In the English system the units are usually in lb/in.^{2} or lb/ft^2. Scientists and engineers usually use the metric system, where pressure is measured in Pascals (Pa). One Pascal is equal to a force of 1 Newton per square meter. A Newton is equal to the force required to accelerate 1 kilogram 1 meter per second per second (1 kg·m/s^2). You can also have kilopascals (kPa), megapascals (mPa), and gigapascals (gPa). Also: 1 Pascal = 10 dyne/cm^2 = 0.01 mbar. 1 atm = 101,325 Pascals = 760 mm Hg = 760 torr (Torricelli barometer) = 14.7 psi. *Note:* psi = pounds per square inch.

62. What is the psi at the bottom of a clarifier, if the water level is 11.11 ft deep?

Equation: $\mathbf{psi} = \dfrac{\text{Depth, ft}}{2.31 \text{ ft/psi}}$

$\text{psi} = \dfrac{11.11 \text{ ft}}{2.31 \text{ ft/psi}} = 4.8095 \text{ psi}$, round to **4.81 psi**

63. A wastewater pond at its deepest point is 15.8 ft deep. What is the psi on the bottom?

Another number to commit to memory is 0.433 psi/ft, which is 1 psi/2.31 ft.

The equation would then be: **psi = (Depth, ft)(0.433 psi/ft)**

$\text{psi} = (15.8 \text{ ft})(0.433 \text{ psi/ft}) = $ **6.84 psi**

HYDRAULIC PRESS CALCULATIONS

Hydraulic press calculations have two fundamental principles: total force equals pressure applied times area the pressure is applied to, and that the force applied to a liquid will be equally distributed within that liquid.

64. **A force of 325 pounds is applied to a small cylinder on a hydraulic jack. The diameter of the small cylinder is 12.0 inches. If the diameter of the large cylinder is 4.50 ft, what is the total lifting force?**

Equation: **Pressure** $= \dfrac{\text{Total force, lb}}{\text{Area, ft}^2}$ for pressure on the small cylinder.

First, convert 12.0 inches to feet: (12.0 in.)(1 ft/12 in.) = 1.00 ft.

$$\text{Pressure} = \frac{325\ \text{lb}}{(0.785)\,(1.00\ \text{ft})\,(1.00\ \text{ft})} = 414.013\ \text{lb/ft}^2$$

Next, calculate the total force on the large cylinder.

Equation: **Total force = (Pressure)(Area)**

Total Force = (414.013 lb/ft²)(0.785)(4.50 ft)(4.50 ft) = 6,581.25 lb, round to **6,580 lb**

65. **A force of 25 pounds is applied to a small cylinder on a hydraulic jack. The diameter of the small cylinder is 6.0 inches. If the diameter of the large cylinder is 3.0 ft, what is the total lifting force?**

Equation: **Pressure** $= \dfrac{\text{Total force, lb}}{\text{Area, ft}^2}$ for pressure on the small cylinder.

First, convert 6.0 inches to feet: (6.0 in.)(1 ft/12 in.) = 0.50 ft.

$$\text{Pressure} = \frac{25\ \text{lb}}{(0.785)\,(0.50\ \text{ft})\,(0.50\ \text{ft})} = 127.39\ \text{lb/ft}^2$$

Next, calculate the total force on the large cylinder.

Equation: **Total force = (Pressure)(Area)**

Total force = (127.389 lb/ft²)(0.785)(3.0 ft)(3.0 ft) = **900 lb**

SCREENING MATERIAL REMOVAL CALCULATIONS

The amount of screening debris should be calculated by operators so that they can plan and properly dispose of the material. A record should be kept each time for the amount of material removed from the screening pits. Screenings are usually disposed of by landfill, incinerated, or ground and returned to the wastewater process. They are very odorous and will attract flies. See the figures in appendix E for placement of wastewater screens.

66. **What is the number of ft³/mil gal of screening material removed at a wastewater treatment plant, if the plant's screenings totaled 107 gallons while processing 3,860,000 gallons?**

First, determine the amount of cubic feet in 107 gallons.

$$\text{Number of ft}^3 = \frac{107 \text{ gal}}{7.48 \text{ gal/ft}^3} = 14.3 \text{ ft}^3$$

Next, convert gallons to mil gal.

$$\text{Number of mil gal} = \frac{3,860,000 \text{ gal}}{1,000,000/\text{mil}} = 3.86 \text{ mil gal}$$

Next, determine the screenings removed in ft³/mil gal.

Equation: **Screenings, ft³/mil gal** $= \dfrac{\textbf{Number of ft}^3}{\textbf{Number of mil gal}}$

$$\text{Screenings, ft}^3/\text{mil gal} = \frac{14.3 \text{ ft}^3}{3.86 \text{ mgd}} = \textbf{3.70 ft}^3\textbf{/mil gal}$$

67. **How many gallons of screenings were removed from a wastewater plant, if the plant processed 2,110,000 gallons and the screenings removed per million gallons was 4.07 ft³/mil gal?**

First, convert gallons to mil gal.

$$\text{Number of mil gal} = \frac{2,110,000 \text{ gal}}{1,000,000/\text{mil}} = 2.11 \text{ mil gal}$$

Next, calculate the number of ft³ of screenings removed by rearranging the following equation:

Equation: **Screenings, ft³/mil gal** $= \dfrac{\text{Number of ft}^3}{\text{Number of mil gal}}$

Rearrange the equation.

Number of ft³ $=$ (Screenings, ft³/mil gal)(Number of mil gal)

Substitute values and solve.

Number of ft³ $=$ (4.07 ft³/mil gal)(2.11 mil gal) $=$ 8.5877 ft³

Lastly, convert the ft³ of screenings removed to gallons of screenings removed.

Screenings removed, gal $=$ (8.5877 ft³)(7.48 gal/ft³) $=$ 64.24 gal, round to **64.2 gal**

SCREENING PIT CAPACITY CALCULATIONS

The operator needs to know the capacity of a screening pit so he or she knows when it should be cleaned based on past records of material removed (above calculations).

68. **How many days will it take to fill a screening pit, if the pit is 5.0 ft by 12 ft and 4.4 ft deep and the average screenings each day total 2.37 ft³?**

First, determine the volume of the pit in ft³.

Pit volume, ft³ $=$ (5.0 ft)(12 ft)(4.4 ft) $=$ 264 ft³

Equation: **Number of days to fill** $= \dfrac{\text{Pit volume, ft}^3}{\text{Screenings removed, ft}^3/\text{day}}$

Substitute values and solve.

Number of days to fill $= \dfrac{264 \text{ ft}^3}{2.37 \text{ ft}^3/\text{days}} =$ 111.39 days, round to **110 days**

69. How many cubic feet is a screening pit, if it would fill in 81.7 days and the average screenings each day are 2.91 ft³?

Equation: **Number of days to fill** $= \dfrac{\text{Pit volume, ft}^3}{\text{Screenings removed, ft}^3/\text{day}}$

Rearrange the equation to solve for pit volume.

Pit volume, ft³ = (Number of days to fill)(Screenings removed, ft³/day)

Substitute values and solve.

Pit volume, ft³ = (81.7 days)(2.91 ft³/day) = 237.747 ft³, round to **238 ft³**

GRIT REMOVAL CALCULATIONS

Grit removal is important for the same reason as screening removal—planning for proper disposal. Grit channels are important in wastewater treatment because by removing the grit from the waste it prevents wear on pumps and deposition in pipelines or channels. It also prevents grit from accumulating in other processes such as digesters or biological contactors. Not all wastewater treatment plants have grit channels and they are not always placed after screens or comminutors. See the figures in Appendix E for where grit channels are commonly placed in different treatment plants.

70. A wastewater plant removes 64 gallons of grit during the processing of 2.65 mil gal. What is the ft³/mil gal removal rate during this interval?

Equation: **Grit removal, ft³/mil gal** $= \dfrac{\text{Number of gallons removed}}{(7.48 \text{ gal/ft}^3)(\text{mil gal treated})}$

Grit removal, ft³/mil gal $= \dfrac{64 \text{ gal}}{(7.48 \text{ gal/ft}^3)(2.65 \text{ mil gal})} = $ **3.23 ft³/mil gal**

71. **How many mil gal of waste was treated by a plant, if the number of gallons of grit removed was 42.5 gallons and the grit removal rate was 1.96 ft³/mil gal?**

Equation: **Grit removal, ft³/mil gal** $= \dfrac{\text{Number of gallons removed}}{(\text{mil gal treated})\,(7.48\ \text{gal/ft}^3)}$

Rearrange to solve for mil gal treated.

Mil gal treated $= \dfrac{\text{Number of gallons removed}}{(\text{Grit removal, ft}^3/\text{mil gal})\,(7.48\ \text{gal/ft}^3)}$

Substitute values and solve.

Mil gal treated $= \dfrac{42.5\ \text{gal}}{(1.96\ \text{ft}^3/\text{mil gal})\,(7.48\ \text{gal/ft}^3)} = 2.8989$ mil gal, round to **2.90 mil gal**

RECIRCULATION RATIO PROBLEMS

The recirculation ratio calculation is used to help operators keep flow variations to a trickling filter to a minimum. This ratio can also help solve process problems such as increasing the rate of hydraulic loading when needed, reducing odors and filter flies, or preventing the trickling filter from drying out during low flows. Recirculation is from the secondary clarifier or trickling filter effluent. The ratio usually ranges from 1:1 to 2:1.

72. **Given the following data, determine the recirculation ratio:**

Influent flow = 1,330,000 gpd
Effluent flow = 1.74 mgd

First, convert gpd to mgd.

Note: The problem can also be solved by converting mgd to gpd since all that is asked for is a ratio.

Number of mgd = (1,330,000 gpd) / (1,000,000 gal/mil) = 1.33 mgd

Equation: **Ratio** $= \dfrac{\text{Recirculated flow}}{\text{Plant influent flow}}$

Ratio = 1.74 mgd/1.33 mgd = 1.308, round to **1.31**

73. **What must have been the trickling filter's effluent flow in mgd, if the influent flow to the wastewater plant was 950,000 gpd and the recirculation ratio was 1.46?**

First, convert gpd to mgd.

Number of mgd = (950,000 gpd) / (1,000,000 gal/mil) = 0.95 mgd

Equation: $\textbf{Ratio} = \dfrac{\text{Recirculated flow}}{\text{Plant influent flow}}$

Rearrange the equation to solve for the recirculation flow (return of the trickling filter's effluent).

Recirculated flow = (Ratio)(Plant influent flow)

Substitute values and solve.

Recirculated flow = (1.46)(0.95) = 1.387 mgd, round to **1.4 mgd**

DETENTION TIME CALCULATIONS

Detention time is simply the time period that starts when wastewater flows into a basin or tank and ends when it flows out of the basin or tank. Detention time is usually calculated for wastewater ponds, oxidation (aerobic) ditches, and clarifiers. Detention times are theoretical, because basins begin to fill with settled sludge and other debris. This causes the true detention time to constantly change (decrease). While it is true that sludge removals will cause the detention time to increase, the true detention time will always be less than theoretical. Also, flows through a basin are never perfectly laminar and thus cause a further decrease in the true detention time. See Figures 5 and 6 in Appendix E for two types of wastewater plants using ponds or Figure 8 in Appendix E for a wastewater plant using an oxidation ditch.

74. **What is the detention time, in hours, if an oxidation ditch has an influent flow of 0.232 mgd and the volume of the oxidation ditch is 28,100 ft³?**

First, convert mgd to gpd.

Number of gpd = (0.232 mgd)(1,000,000 gal/mil) = 232,000 gpd

Next, convert the volume of the oxidation ditch from ft^3 to gallons.

Oxidation ditch volume, gal = (28,100 ft^3)(7.48 gal/ft^3) = 210,188 gal

Lastly, calculate the detention time in hours.

Equation: **Detention time, hr** $= \dfrac{(\text{Volume, gal})\,(24\ \text{hr/day})}{\text{Flow, gpd}}$

Detention time, hr $= \dfrac{(210{,}188\ \text{gal})\,(24\ \text{hr/day})}{232{,}000\ \text{gpd}} = 21.74$ hr, round to **21.7 hr**

75. **What is the detention time in days for a wastewater treatment pond given the following data?**

Pond = averages 365 ft by 171 ft by 7.25 ft
Flow = 43,500 gpd

First, calculate the volume of the waste pond in gallons.

Volume, gal = (365 ft)(171 ft)(7.25 ft)(7.48 gal/ft^3) = 3,384,765 gal

Equation: **Detention time, days** $= \dfrac{(\text{Volume, gal})}{\text{Flow, gpd}}$

Detention time, days $= \dfrac{(3{,}384{,}765\ \text{gal})}{43{,}500\ \text{gpd}} = 77.81$ days, round to **77.8 days**

76. **What is the detention time in days for a wastewater treatment pond given the following data?**

Pond = averages 315 ft by 122 ft by 7.45 ft
Flow = 4.25 ft³/min

First, convert ft³/s to gpd.

Number of gpd = (4.25 ft³/min)(1,440 min/day)(7.48 gal/ft³) = 45,778 gpd

Next, calculate the volume of the waste pond in gallons.

Volume, gal = (315 ft)(122 ft)(7.45 ft)(7.48 gal/ft³) = 2,141,550 gal

Equation: **Detention time, days** $= \dfrac{\text{Volume, gal}}{\text{Flow, gpd}}$

Detention time, days $= \dfrac{2,141,550 \text{ gal}}{45,778 \text{ gpd}} = 46.78$ days, round to **46.8 days**

WEIR OVERFLOW RATE PROBLEMS

A weir is like a small dam, gate, notch, or other barrier placed across a basin to help regulate water out of the basin. The weir overflow rate is used to determine the velocity of wastewater over the weir. The velocity informs the operator about the efficiency of the sedimentation process. At constant wastewater flow, the shorter the length of the weir, the faster the water velocity will be out of the basin. Conversely, the longer the weir length, the slower the velocity will be out of the basin. See Figures 1, 2, 3, 7, and 8 in Appendix E for five types of wastewater plants using a clarifier.

77. **A circular clarifier is 69.8 ft in diameter. If flow to the clarifier averages 1.36 mgd, what is the weir overflow rate in gpd/ft?**

First, convert mgd to gpd.

Number of gpd = (1.36 mgd)(1,000,000 gal/mil) = 1,360,000 gpd

Next, calculate the length of the weir.

Weir length, ft = π(Diameter, ft)

Weir length, ft = Circumference = (3.14)(69.8 ft) = 219.172 ft

Next, solve for the weir overflow rate.

Equation: **Weir overflow rate, gpd/ft** $= \dfrac{\text{Flow, gpd}}{\text{Weir length, ft}}$

Weir overflow rate, gpd/ft $= \dfrac{1{,}360{,}000 \text{ gpd}}{219.172 \text{ ft}}$ = 6205.17 gpd/ft, round to **6,210 gpd/ft**

78. **What is the weir overflow rate in gpd/ft, if the flow is 0.475 mgd and the radius of the clarifier is 29.75 ft?**

First, calculate the length of the weir.

Weir length, ft = 2π(radius, ft)

Weir length, ft = 2(3.14)(29.75 ft) = 186.83 ft

Next, convert mgd to gpd.

Number of gpd = (0.475 mgd)(1,000,000 gal/mil) = 475,000 gpd

Next, solve for the weir overflow rate.

Equation: **Weir overflow rate, gpd/ft** $= \dfrac{\text{Flow, gpd}}{\text{Weir length, ft}}$

Weir overflow rate, gpd/ft $= \dfrac{475{,}000 \text{ gpd}}{186.83 \text{ ft}}$ = 2,542.42 gpd/ft, round to **2,540 gpd/ft**

SURFACE OVERFLOW RATE CALCULATIONS

Surface overflow rate determinations are used to determine the loading on clarifiers. The flow amount used in these calculations only counts plant flow, not recirculation.

79. What is the surface overflow rate in gpd/ft², if the clarifier is 30.2 ft in radius and the flow into the basin is 485 gpm?

First, determine the area of the clarifier.

Area $= \pi r^2$ where $\pi = 3.14$

Area $= (3.14)(30.2 \text{ ft})(30.2 \text{ ft}) = 2{,}863.81 \text{ ft}^2$

Next, convert gpm to gpd.

Number of gpd $= (485 \text{ gpm})(1{,}440 \text{ min/day}) = 698{,}400$

Lastly, calculate the surface overflow rate.

Equation: **Surface overflow rate** $= \dfrac{\text{Flow, gpd}}{\text{Area, ft}^2}$

Surface overflow rate $= \dfrac{698{,}400 \text{ gpd}}{2{,}863.81 \text{ ft}^2} = 243.87 \text{ gpd/ft}^2$, round to **244 gpd/ft²**

80. Calculate the surface overflow rate in gpd/ft² for a clarifier that has a diameter of 48.5 ft and the flow is 0.730 mgd.

First, determine the area of the clarifier.

Area $= \pi r^2$ where r $=$ Diameter/2 $= 48.5 \text{ ft}/2 = 24.25 \text{ ft}$

Area $= (3.14)(24.25 \text{ ft})(24.25 \text{ ft}) = 1{,}846.52 \text{ ft}^2$

Next, convert mgd to gpd/ft².

Number of gpd/ft^2 = (0.730 mgd)(1,000,000 gal/mil) = 730,000 gal

Now, calculate the surface overflow rate.

Equation: **Surface overflow rate** $= \dfrac{\text{Flow, gpd}}{\text{Area, ft}^2}$

Surface overflow rate $= \dfrac{730,000 \text{ gpd}}{1,846.52 \text{ ft}^2} = 395.34$ gpd/ft^2, round to **395 gpd/ft^2**

FLOW AND VELOCITY CALCULATIONS

Operators need to know the flow and velocity of the wastewater throughout the different plant processes, for example to feed proper dosages of chemicals to treat wastewaters, to know how many clarifiers or ponds to use or how much supernatant to recirculate, and for settling purposes, among other uses.

81. **What is the flow in a rectangular channel in feet per second (ft/s), given the following data?**

Channel = 5.75 ft wide
Depth of wastewater flow = 3.42 ft
Flow = 1,892 gpm

Equation: **Velocity, ft/s** $= \dfrac{\text{Flow, gpm}}{(\text{Width, ft})\,(\text{Depth, ft})\,(60 \text{ sec/min})\,(7.48 \text{ gal/ft}^3)}$

Substitute values and solve.

Velocity, ft/s $= \dfrac{1,892 \text{ gpm}}{(5.75 \text{ ft})\,(3.42 \text{ ft})\,(60 \text{ sec}/\text{min})\,(7.48 \text{ gal/ft}^3)} =$ **0.214 ft/s**

82. **If a 3.00-ft diameter chemical tank drops 4.20 inches in exactly 2 hours, what is the pumping rate for the chemical in gpm?**

First, determine the amount in feet the tank level dropped.

Drop, ft = (4.20 in.)(1 ft/12 in.) = 0.350 ft

Then, determine the volume in gallons for the drop in level of the tank.

Equation: **Volume, gal = (0.785)(Diameter)²(Drop, ft)(7.48 gal/ft³)**

Substitute values and solve.

Volume, gal = (0.785)(3.00 ft)(3.00 ft)(0.350 ft)(7.48 gal/ft³) = 18.496 gal

Next, convert 2 hours to minutes.

Number of minutes = (2 hr)(60 min/hr) = 120 min

Now, calculate the pumping rate in gpm.

Equation: **Pumping rate = Flow, gal/Time, min**

Pumping rate = 18.496 gal/120 min = **0.154 gpm**

83. A 10.0-inch pipeline is flowing full and at 555 gpm. What is the velocity in ft/s?

First, convert the diameter of the pipe in inches to feet.

Number of ft = (10.0 in.)(1 ft/12 in.) = 0.833 ft

Next, determine the cross-sectional area of the pipe, which is defined with the following equation.

Equation: **Area, ft² = πr²**, where radius = Diameter/2, thus 0.8333 ft/2 = radius = 0.4167 ft

Area, ft² = (3.14)(0.4167 ft)(0.4167 ft) = 0.545 ft²

Next, calculate the velocity in ft/s using the following equation.

Equation: **Velocity, ft/s** $= \dfrac{\text{Flow, gpm}}{(\text{Area, ft}^2)\,(60 \text{ sec/min})\,(7.48 \text{ gal/ft}^3)}$

Substitute values and solve.

Velocity, ft/s $= \dfrac{555 \text{ gpm}}{(5.45 \text{ ft}^2)\,(60 \text{ sec}/\text{min})\,(7.48 \text{ gal/ft}^3)} = 0.2269$ ft/s, round to **0.227 ft/s**

84. If a 25.0-ft diameter clarifier drops 18.87 inches in exactly 6 hours, what is the pumping rate out of the tank in gpm?

First, determine the amount in feet the tank level dropped.

Drop, ft = (18.87 in.)(1 ft/12 in.) = 1.5725 ft

Then, determine the volume in gallons for the drop in level of the tank.

Equation: **Volume, gal = (0.785)(Diameter)²(Drop, ft)(7.48 gal/ft³)**

Substitute values and solve.

Volume, gal = (0.785)(25.0 ft)(25.0 ft)(1.5725 ft)(7.48 gal/ft³) = 5,770.88 gal

Next, convert 6 hours to minutes.

Number of minutes = (6 hr)(60 min/hr) = 360 min

Now, calculate the pumping rate in gpm.

Equation: **Pumping rate = Flow, gal/Time, min**

Pumping rate = 5,770.88 gal/360 min = 16.03 gpm, round to **16.0 gpm**

The following two problems involve flow through a pipeline that is **not** flowing full. The calculations are almost the same as determining flow in a full pipeline, except the multiplication factor of 0.785 is not used. A new factor is used and is based on the liquid level in the pipe divided by the pipe's diameter. These factors are presented in the depth/Diameter table in Appendix D.

Note: Answers are only approximate.

85. **A 24-inch sewage pipeline is flowing at a velocity of 1.45 ft/s and the depth of the sewage averages 9.3 inches. Determine the flow in the pipeline in ft³/s.**

First, divide the depth of sewage flow by the diameter of the pipe. Converting inches to feet is not necessary.

Ratio = depth/Diameter = 9.3 in./24 in. = 0.3875, round to 0.39

Note: Extrapolation can also be used if more accuracy is required.

Next, determine the factor that needs to be used.

In Appendix D look up 0.39 under the column d/D. The number immediately to the right will be the factor that needs to be used. In this case it is 0.2836. This will be the number used rather than 0.785.

Next, convert the pipe's diameter from inches to feet.

$$\text{Number of feet} = \frac{24 \text{ in.}}{12 \text{ in./ft}} = 2.0$$

Equation: **Flow, ft³/sec = (Area, ft²)(Velocity, ft/s)**

Where the area = (Factor)(Diameter)²

Substitute values and solve.

Flow, ft³/s = (0.2836)(2.0 ft)(2.0 ft)(1.45 ft/s) = 1.645 ft³/s, round to **1.6 ft³/s**

86. **A 36-inch sewage pipeline is flowing at a velocity of 1.18 ft/s and the depth of the sewage averages 10.4 inches. Determine the flow in the pipeline in gpm.**

First, divide the depth of sewage flow by the diameter of the pipe. Converting inches to feet is not necessary.

Ratio = depth/Diameter = 10.4 in./36 in. = 0.2889, round to 0.29

Note: Again, extrapolation can also be used if more accuracy is required.

Next, determine the factor that needs to be used.

In Appendix D, look up 0.29 under the column d/D. The number immediately to the right will be the factor that needs to be used. In this case it is 0.1890. This will be the number used rather than 0.785.

Next, convert the pipe's diameter from inches to feet.

$$\text{Number of feet} = \frac{36 \text{ in.}}{12 \text{ in./ft}} = 3.0$$

Equation: **Flow, ft³/sec = (Area, ft²)(Velocity, ft/s)**

Where the area = (Factor)(Diameter)²

Substitute values and solve.

Flow, ft³/s = (0.1890)(3.0 ft)(3.0 ft)(1.18 ft/s) = 2.007 ft³/s

Now, convert ft³/s to gpm.

Flow, gpm = (2.007 ft³/s)(60 s/min)(7.48 gal/ft³) = 900.74 gpm, round to **900 gpm**

PUMP DISCHARGE PROBLEMS

Operators need to understand pump discharge calculations, for example to plan treatment processes and time, to determine how long a pump will take to discharge a certain amount of wastewater or chemical to treat the wastewater, and maybe to change the size of a pump to fit the need better.

87. **Given the following data, determine the rate a pump discharges into a tank in gpm:**

Duration pump operates = 29 hr and 33 min
Tank diameter = 30.1 ft
Wastewater level at beginning of pumping = 2.85 ft
Wastewater level at end of pumping = 13.44 ft

First, find the number of minutes the pump worked.

Number of min = (29 hr)(60 min/hr) + 33 min = 1,773 min

Next, calculate the change in level during pumping.

Level change, ft = 13.44 ft − 2.85 ft = 10.59 ft

Next, calculate the volume in gallons added to the tank by the pump.

Equation: **Volume, gal = (0.785)(Diameter)^2(Level change, ft)(7.48 gal/ft^3)**

Volume, gal = (0.785)(30.1 ft)(30.1 ft)(10.59 ft)(7.48 gal/ft^3) = 56,337.84 gal

Now, calculate the pump's discharge rate in gpm.

Equation: **Pump's discharge rate, gpm** $= \dfrac{\text{Discharge, gal}}{\text{Time, min}}$

Substitute values and solve.

Pump's discharge rate, gpm $= \dfrac{56,337.84 \text{ gal}}{1,773 \text{ min}} = 31.775$ gpm, round to **32 gpm**

88. **How long will it take in hours and minutes to empty a tanker truck with aluminum sulfate (alum), if the truck's pump unloads the alum at 65 gpm and a total of 17,050 liters needs to be unloaded? The tank's capacity is 10,000 gallons and it already has about 2,000 gallons of alum.**

First, determine the number of gallons in 17,050 liters.

$$\text{Number of gal} = \frac{17,050 \text{ liters}}{3.785 \text{ liters/gal}} = 4,504.62 \text{ gal}$$

Then, divide the number of gallons by the pumping rate.

Time to pump = 4,504.62 gal/65 gpm = 69.3 min

Divide by 60 min/hr.

69.3 min/60 min/hr = 1.155 hr

Next, find how many minutes are in 0.155 hr by multiplying by 60 min/hr.

(0.155 hr)(60 min/hr) = 9.3 min, round to 9 min.

The unloading time = 1 hr and 9 min

89. **How long will it take in hours and minutes for a pump to fill an empty 12.5-ft radius tank to the 12.5-ft level, if the pumping rate is 75 gpm?**

First, calculate the volume in gallons of the tank to the 12.5-ft level.

Equation: **Volume, gal = π(radius)²(Depth, ft)(7.48 gal/ft³)**

Volume, gal = 3.14(12.5 ft)(12.5 ft)(12.5 ft)(7.48 gal/ft³) = 45,873.44 gal

Next, calculate the pump's discharge rate in gpm.

Equation: **Pumping time, min = $\dfrac{\text{Discharge, gal}}{\text{Pump rate, gpm}}$**

Substitute values and solve.

$$\text{Pump's discharge rate, gal} = \frac{45,873.44 \text{ gal}}{75 \text{ gpm}} = 611.6459 \text{ min}$$

Next, divide by 60 (60 min/hr) to determine the number of hours.

Number of hours = (611.6459)/60 min/hr) = 10.1941 hr

Now, determine the number of minutes in 0.1941 hours by multiplying by 60 (60 min/hr).

Number of minutes = (0.1941 hr)(60 min/hr) = 11.646 min, round to 12 min

Thus the tank will be filled to the 12.5-ft level in **10 hours and 12 minutes.**

SOLIDS AND HYDRAULIC LOADING RATE CALCULATIONS

Solids and hydraulic loading rate calculations are used to determine the solids or hydraulic loading on clarifiers, trickling filters, and other processes. These calculations are important to know so operators for example can determine when to discharge sludge from a clarifier, or to know the contact time between organisms in a trickling filter and the food entering that trickling filter.

90. **A trickling filter has a diameter of 109.5 ft. If the flow through the filter is 2.71 mgd and the recirculation rate is 25% of the flow rate, what is the hydraulic loading rate on a trickling filter in gallons per day per square foot (gpd/ft²)?**

First, determine the total flow in gallons per day (gpd) through the trickling filter.

Total flow, gal = [2.71 mgd + 2.71 mgd(25%/100%)](1,000,000/mil)

Total flow, gal = [2.71 mgd + 0.6775 mgd](1,000,000/mil) = 3,387,500 gpd

Next, determine the surface area in ft² for the clarifier.

Area $= \pi r^2$ where r = Diameter/2 = 109.5 ft/2 = 54.75 ft

Clarifier surface area, ft² = (3.14)(54.75 ft)(54.75 ft) = 9,412.35 ft²

Lastly, calculate the hydraulic loading rate.

Hydraulic loading rate $= \dfrac{\text{Total flow, gpd}}{\text{Surface area, ft}^2}$

Hydraulic loading rate $= \dfrac{3,387,500 \text{ gpd}}{9,412.35 \text{ ft}^2} = 359.899$ gpd/ft², round to **360 gpd/ft²**

91. **Calculate the solids loading rate on a secondary clarifier with a diameter of 68.5 ft, flow rate of 1,350,000 gpd, and a mixed liquor suspended solids (MLSS) of 3,425 mg/L.**

First, determine the area of the clarifier.

Area $= \pi r^2$ where r = Diameter/2 = 68.5 ft/2 = 34.25 ft

Area = (3.14)(34.25 ft)² = 3,683.42 ft²

Next, convert gpd to mgd.

Number of mgd = (1,350,000 gpd) / (1,000,000 gal/mil) = 1.35 mgd

Finally, calculate the solids loading rate.

Equation: **Solids loading rate** $= \dfrac{(\text{MLSS, mg/L})(\text{mgd})(8.34 \text{ lb/gal})}{\text{Area, ft}^2}$

Solids loading rate $= \dfrac{(3,425 \text{ mg/L})(1.35 \text{ mgd})(8.34 \text{ lb/gal})}{3,683.42 \text{ ft}^2}$

Solids loading rate = 10.469 lb of solids/d/ft², round to **10.5 lb of solids/d/ft²**

92. **What is the solids loading rate in lb of solids/d/ft², given the following data?**

Secondary clarifier radius = 34.95 ft
Primary effluent flow = 1.22 mgd
Return of activated sludge is 0.65 mgd
Mixed liquor suspended solids (MLSS) = 3,060 mg/L
Specific gravity (sp gr) of the solids is 1.03

First, determine the total flow.

Total flow = Primary flow + Return of activated sludge

Total flow = 1.22 mgd + 0.65 mgd = 1.87 mgd

Next, calculate the area of the clarifier.

Area = πr^2

Area = (3.14)(34.95 ft)² = 3,835.518 ft²

Next, determine the lb/gal of solids.

Solids, lb/gal = (8.34 lb/gal)(1.03 sp gr) = 8.59

Next, calculate the solids loading rate.

Equation: **Solids loading rate** $= \dfrac{(\text{MLSS, mg/L}) (\text{mgd}) (8.34 \text{ lb/gal})}{\text{Area, ft}^2}$

Solids loading rate $= \dfrac{(3,060 \text{ mg/L}) (1.87 \text{ mgd}) (8.59 \text{ lb/gal})}{3,835.518 \text{ ft}^2}$

Solids loading rate = 12.815 lb of solids/d/ft², round to **13 lb of solids/d/ft²**

93. What is the hydraulic loading rate for a pond that is 17.3 acre-ft in gallons per day per ft² (gpd/ft²), if the flow into the pond is 4.06 mgd?

Since the problem asks for gpd, first convert the volume of the pond in acre-ft to gallons.

Know from appendix A: 1 acre-ft = 43,560 ft²

Area of pond, ft² = (17.3 acre-ft)(43,560 ft²/acre-ft) = 753,588 ft²

Next, convert mgd to gallons.

Flow into pond, gal = (4.06 mgd)(1,000,000/mil) = 4,060,000 gpd

Lastly, divide the flow.

Equation: **Hydraulic loading rate** $= \dfrac{\text{Total flow, gpd}}{\text{Surface area, ft}^2}$

Hydraulic loading rate $= \dfrac{4,060,000 \text{ gpd}}{753,588, \text{ ft}^2} = 5.388$ gpd/ft², round to **5.39 gpd/ft²**

94. If a 60.3-ft diameter gravity thickener has 29,500 gpd of biosolids added to it and the biosolids contain 1.95% solids, what is the solids loading rate in lb/d/ft²?

First, determine the surface area of the gravity thickener.

Surface area of gravity thickener, ft² = (0.785)(60.3 ft)(60.3 ft) = 2,854.33 ft²

Now, calculate the solids loading rate.

Equation: **Solids loading rate, lb/d/ft²** $= \dfrac{(\text{Percent solids})\,(\text{Biosolids added, gpd})\,(8.34 \text{ lb/gal})}{(\text{Surface area, ft}^2)}$

Solids loading rate, lb/d/ft² $= \dfrac{(1.95\%/100\%)\,(29,500 \text{ gpd})\,(8.34 \text{ lb/gal})}{(2,854.33 \text{ ft}^2)}$

Substitute values and solve.

Solids loading rate, lb/d/ft² = **1.68 lb/d/ft²**

95. What is the solids loading rate for a secondary clarifier given the following information:

Radius of clarifier = 70.5 ft
Primary effluent flow = 4,225,000 gpd
Return sludge flow = 1,025,000 gpd
Mixed Liquor Suspended Solids (MLSS) = 2,080 mg/L

First, determine the surface area of the clarifier.

Know: **Area** = πr^2 where r equals the radius

Clarifier surface area, ft² = (3.14)(70.5 ft)(70.5 ft) = 15,606.585 ft²

Next, determine the total flow in mgd.

$$\text{Total clarifier flow} = \frac{4,225,000 \text{ gpd} + 1,025,000 \text{ gpd}}{1,000,000/\text{mil}} = 5.25 \text{ mgd}$$

Lastly, determine the solids loading rate of MLSS.

$$\textbf{Solids loading rate, lb MLSS/d/ft}^2 = \frac{(\text{MLSS, mg/L})(\text{mgd})(8.34 \text{ lb/gal})}{\text{Surface area, ft}^2}$$

$$\text{Solids loading rate, lb MLSS/d/ft}^2 = \frac{(2,080 \text{ mg/L, MLSS})(5.25 \text{ mgd})(8.34 \text{ lb/gal})}{15,606.585 \text{ ft}^2}$$

Solids loading rate, lb MLSS/d/ft² = **5.84 lb MLSS/d/ft²**

SLUDGE PUMPING PROBLEMS

Sludge pumping calculations are important for operators to determine so they know how much sludge and solids are being loaded into a digester to prevent underloading or overloading of the digester. Also, operators need to know how much sludge is being pumped to other sludge processing applications such as sludge thickening, filter presses, or land application.

96. **Given the following data, calculate the lb/day of sludge pumped from a clarifier to a sludge thickener:**

Pump operates exactly 15 minutes every 2.00 hours
Pump rate = 18.5 gpm
Solids content = 4.87%
Specific gravity (sp gr) = 1.04

Equation: **Sludge, lb/day = (Pumping, min/day)(24 hr/day)(Pump rate, gpm)(8.34 lb/gal)(sp gr of sludge)**

Sludge, lb/day = (15 min/2.0 hr)(24 hr/day)(18.5 gpm)(8.34 lb/gal)(1.04 sp gr)

Sludge, lb/day = 28,883.088 lb/day, round to **28,900 lb/day of Sludge**

97. **Given the following data calculate the lb/day of solids pumped to a sludge thickener:**

Sludge sample = 2,011.230 grams (g)
Solids content after drying = 79.106 g
Pump operates exactly 10 minutes every 1.00 hours
Pump rate = 16.5 gpm
Specific gravity (sp gr) = 1.04
Clarifier effluent flow = 1.48 mgd

First, determine the percent solids in the sludge.

Equation: **Percent solids = (Dry solids, g)(100%)/Sludge sample, g**

Percent Solids = (79.106 g)(100%)/2,011.230 g = 3.933% solids

Now, calculate the solids pumped in lb/day.

Equation: **Solids, lb/day = (Pumping, min/day)(24 hr/day)(Pump rate, gpm) (Percent solids)(8.34 lb/gal)(sp gr, sludge)**

Solids, lb/day = (10 min/hr)(24 hr/day)(16.5 gpm)(3.933%/100 %)(8.34 lb/gal)(1.04 sp gr)

Solids, lb/day = 1,350.89 lb/day, round to **1,350 lb/day of Solids**

98. **How many lb/day of solids were pumped to a digester, if a sludge pump operates exactly 10 minutes every hour at a rate of 24.5 gpm, the percent solids in the sludge was 4.83%, and the specific gravity of the sludge was 1.06?**

Solids, lb/day = (Pumping, min/day)(24 hr/day)(Pump rate, gpm)(8.34 lb/gal) (sp gr, sludge)

Solids, lb/day = (10 min/hr)(24 hr/day)(24.5 gpm)(8.34 lb/gal)(1.06 sp gr)(4.83%/100 %)

Solids, lb/day = 2,510.71 lb/day, round to **2,510 lb/day of Solids**

BIOSOLIDS PUMPING AND PRODUCTION PROBLEMS

Biosolids pumping calculations provide operators accurate process control data for the sedimentation process. Biosolids are mostly composed of water with the biosolids ranging from only 3 to 7% by volume.

99. **What is the biosolids production in lb/mil gal, if a wastewater treatment plant produces 174,000 gallons of biosolids in a 30-day month and the plant treated 2.07 mgd on average?**

Equation: **Biosolids, lb/mil gal** $= \dfrac{\text{(Biosolids, gal/day)}(8.34\ \text{lb/gal})}{\text{(Flow, mgd)}(\text{Number of days})}$

Biosolids, lb/mil gal $= \dfrac{(174{,}000\ \text{gal/day})(8.34\ \text{lb/gal})}{(2.07\ \text{mgd})(30\ \text{days})}$

Biosolids, lb/mil gal = 23,368 lb/mil gal, round to **23,400 lb/mil gal Biosolids**

100. **If the plant flow averages 3.77 mgd and production of biosolids averages 16,880 gal/day, what is the biosolids production in wet tons per year?**

Equation: **Biosolids, wet tons/yr** $= \dfrac{(\text{Biosolids, lb/mil gal})\,(\text{mgd})\,(365\,\text{days/yr})}{2,000\,\text{lb/wet ton}}$

Substitute values and solve.

Biosolids, wet tons/yr $= \dfrac{(16,880\,\text{lb/mil gal})\,(3.77\,\text{mgd})\,(365\,\text{days/yr})}{2,000\,\text{lb/wet ton}}$

Biosolids, wet tons/yr $=$ 11,614 wet tons/yr, round to **11,600 Wet tons/yr Biosolids**

101. **What is the estimated biosolids pumping rate for the following system?**

Plant flow = 4.92 mgd
Removed biosolids = 1.11%
Influent total suspended solids (TSS) = 282 mg/L
Effluent TSS = 118 mg/L

Equation: **Estimated pumping rate =**

$$\frac{(\text{Influent TSS, mg/L} - \text{Effluent TSS, mg/L})\,(\text{Flow, mgd})\,(8.34\,\text{lb/gal})}{(\text{Percent solids in sludge})\,(\text{Sludge, lb/gal})\,(1,440\,\text{min/day})}$$

Substitute values and solve.

Estimated pumping rate $= \dfrac{(282\,\text{TSS mg/L} - 118\,\text{TSS, mg/L})\,(4.92\,\text{mgd})\,(8.34\,\text{lb/gal})}{(1.11\%/100\%)\,(8.34\,\text{lb/gal})\,(1,440\,\text{min/day})}$

Estimated pumping rate $= \dfrac{(164\,\text{TSS mg/L})\,(4.92\,\text{mgd})\,(8.34\,\text{lb/gal})}{(1.11\%/100\%)\,(8.34\,\text{lb/gal})\,(1,440\,\text{min/day})}$

Estimated pumping rate = **50.5 gpm**

102. What is the estimated biosolids pumping rate for the following system?

Plant flow $= 2.28$ mgd
Removed biosolids $= 1.41\%$
Influent total suspended solids (TSS) $= 278$ mg/L
Effluent TSS $= 109$ mg/L
Biosolids weight $= 8.45$ lb/gal

Equation: **Estimated pumping rate =**

$$\frac{(\text{Influent TSS, mg/L} - \text{Effluent TSS, mg/L})(\text{Flow, mgd})(8.34\ \text{lb/gal})}{(\text{Percent solids in sludge})(\text{Sludge, lb/gal})(1{,}440\ \text{min/day})}$$

Substitute values and solve.

$$\text{Estimated pumping rate} = \frac{(278\ \text{TSS mg/L} - 109\ \text{TSS, mg/L})(2.28\ \text{mgd})(8.34\ \text{lb/gal})}{(1.41\%/100\%)(8.45\ \text{lb/gal})(1{,}440\ \text{min/day})}$$

$$\text{Estimated pumping rate} = \frac{(169\ \text{TSS mg/L})(2.28\ \text{mgd})(8.34\ \text{lb/gal})}{(1.41\%/100\%)(8.45\ \text{lb/gal})(1{,}440\ \text{min/day})}$$

Estimated pumping rate $=$ **18.7 gpm**

103. Given the following data, calculate the amount of solids and volatile solids removed in lb/day:

Pumping rate $= 185$ gpm
Pump frequency $= 12$ times/day
Pumping cycle $= 15$ minutes exactly per cycle
Solids $= 3.03\%$
Volatile solids (VS) $= 59.4\%$

First, determine the solids removal in lb/day.

Equation: **Solids, lb/day =**

(Time, min/cycle)(cycles/day)(Pump rate, gpm)(8.34 lb/gal)(Percent solids)

Substitute values and solve.

Solids, lb/day = (15 min/cycle)(12 cycles/day)(185 gpm)(8.34 lb/gal)(3.03%/100%)

Solids, lb/day = 8,414.98 lb/day, round to **8,410 lb/day Solids**

Next, calculate the amount of volatile solids removed in lb/day.

Equation: **VS, lb/day =**

(Time, min/cycle)(cycles/day)(Pump rate, gpm)(8.34 lb/gal)(Percent, solids)(Percent VS)

Substitute values and solve.

VS, lb/day = (15 min/cycle)(12 cycles/day)(185 gpm)(8.34 lb/gal)(3.03%/100%)(59.4%/100%)

VS, lb/day = 4,998.496 lb/day, round to **5,000 lb/day VS**

104. **What is the concentration factor (CF) of biosolids, if the influent biosolids to a gravity thickener is 1.85% and the effluent biosolids is 5.15%?**

Equation: $\mathbf{CF = \dfrac{Percent\ thickened\ biosolids}{Percent\ influent\ biosolids}}$

Substitute values and solve.

$CF = \dfrac{5.15\%}{1.85\%} = \mathbf{2.78}$

WASTE ACTIVATED SLUDGE PUMPING RATE CALCULATIONS

These calculations are used as a planning tool by the operator. The waste activated sludge (WAS) suspended solids (SS) are pumped out of the secondary clarifier and wasted or returned to the aeration tank. It is better to pump continuously rather than intermittently and not to change the amount by more than 15% from one day to the next.

105. **Determine the waste activated sludge (WAS) pumping rate in gpm, if 3,380 lb/ day are to be wasted and the WAS suspended solids concentrations 3,210 mg/L.**

First, use the "pounds" equation to solve for the number of mgd.

Equation: **WAS, lb/day = (WAS, mg/L)(Number of mgd)(8.34 lb/gal)**

Rearrange the equation to solve for mg/L.

$$\text{Number of mgd} = \frac{(\text{WAS, lb/day})}{(\text{Number of mg/L WAS})(8.34 \text{ lb/gal})}$$

Substitute values and solve.

$$\text{Number of mgd} = \frac{(3,380 \text{ lb/day})}{(3,210 \text{ mg/L WAS})(8.34 \text{ lb/gal})} = 0.12625 \text{ mgd}$$

Now, convert mgd to gpm.

$$\text{Number of gpm} = \frac{(0.12625 \text{ mgd})(1,000,000/\text{mil})}{(1,440 \text{ min}/\text{day})} = \textbf{87.7 gpm}$$

106. **Determine the waste activated sludge (WAS) pumping rate in gpm given the following data:**

Amount of WAS to be wasted = 4,090 lb/day
WAS suspended solids concentrations = 3,475 mg/L

First, use the "pounds" equation to solve for the number of mgd.

Equation: **WAS, lb/day = (WAS, mg/L)(Number of mgd)(8.34 lb/gal)**

Rearrange the equation to solve for mg/L.

$$\text{Number of mgd} = \frac{(\text{WAS}, \text{lb/day})}{(\text{Number of mg/L WAS})(8.34\,\text{lb/gal})}$$

Substitute values and solve.

$$\text{Number of mgd} = \frac{(4,090\,\text{lb/day})}{(3,475\,\text{mg/L WAS})(8.34\,\text{lb/gal})} = 0.1411\,\text{mgd}$$

Now, convert mgd to gpm.

$$\text{Number of gpm} = \frac{(0.1411\,\text{mgd})(1,000,000/\text{mil})}{(1,440\,\text{min/day})} = \textbf{98.0 gpm}$$

TOTAL HEAD, TOTAL STATIC HEAD, AND HEAD LOSS CALCULATIONS

These calculations tell the operator how hard a pump has to work against the static head (the height above the pump that the liquid is discharged to), as well as head losses caused from friction from the water and pipe contact, bends in the pipe, and appurtenances (valves or orifices) along the pipeline. For simplicity, this book will refer to total dynamic head (TDH in some literature) as total head.

107. **Water is being pumped from a water source with an elevation of 3,895 ft to an elevation of 4,140 ft. What is the total head, if friction and minor head losses are 21.8 ft?**

Equation: **Total head, ft = Total static head, ft (difference in elevation) + Head losses, ft**

Total head, ft = (4,140 ft − 3,895 ft) + 21.8 ft = 245 + 21.8 ft = 266.8 ft, round to **267 ft**

108. **If water is being pumped from a water source with an elevation of 68 ft to an elevation of 141 ft and the total dynamic head was 84.5, what must have been the amount of friction and minor head losses?**

Equation: **Total head, ft = Total static head, ft (difference in elevation) + Head losses, ft**

Rearrange the equation to solve for head losses.

Head losses, ft = Total head, ft − Total static head, ft (difference in elevation)

Head losses, ft = 84.5 ft − (141 ft − 68 ft) = 84.5 ft − 73 ft = **11.5 ft**

109. **Given the following data, determine the elevation of a pump and its source:**

Friction loss and minor head losses = 15.8 ft
Total dynamic head (TDH) = 467 ft?
Water destination = 3,544 ft

Equation slightly modified for clarity (showing that the destination elevation minus the source elevation is the total static head):

Total head, ft = (Destination elevation, ft − Source elevation, ft) + Head losses, ft

Rearrange the equation to solve for the source elevation.

Source elevation, ft = Destination elevation, ft + Head losses, ft − Total head, ft

Substitute values and solve.

Source elevation, ft = 3,544 ft + 15.8 ft − 467 ft = 3,092.8 ft, round to **3,090 ft**

PUMPING HORSEPOWER, EFFICIENCY, AND COSTING CALCULATIONS

These types of calculations can be used for determining pump size, efficiency, and costing.

110. Find the motor horsepower (mhp) for a pump with the following parameters:

Motor efficiency (ME): 91.5%
Total Head (TH): 306 ft
Pump efficiency (PE): 78.2%
Flow: 7.85 mgd

First, convert mgd to gpm.

Gpm = (7.85 mgd)(1,000,000/mil)(1 day/1,440 min) = 5,451.4 gpm

The equation for determining the mhp with the given data is different then the problem above.

Equation: $\mathbf{mhp} = \dfrac{(\text{Flow, gpm})(\text{TH, ft})}{(3,960)(\text{ME})(\text{PE})}$

Substitute values and solve.

$mhp = \dfrac{(5,451.4 \text{ gpm})(306 \text{ ft})}{(3,960)(91.5\%/100\% \text{ ME})(78.2\%/100\% \text{ PE})}$

mhp = 588.7 mhp, round to **589 mhp**

111. What is the water horsepower (whp) for a pump that has a motor efficiency (ME) of 92.5%, pump efficiency (PE) of 81%, and a motor horsepower (mhp) of 250?

Equation: **Water horsepower = (mhp)(ME)(PE)**

Water horsepower = (250 mhp)(92.5%/100% ME)(81%/100% PE)

Water horsepower = 187 whp, round to **190 whp**

112. **What is the brake horsepower (bhp) for a pump, if the water horsepower (whp) is 120 and the pump efficiency (PE) is 79%?**

Equation: **Brake horsepower = whp/PE**

$$\text{Brake hp} = \frac{120 \text{ whp}}{79\%/100\% \text{ PE}} = 151.9 \text{ bhp, round to } \textbf{150 bhp}$$

113. **What is the motor horsepower (mhp), if the bhp is 88 and the motor efficiency is 89%?**

Equation: **Motor hp = bhp/ME**

$$\text{Motor horsepower} = \frac{88 \text{ bhp}}{89\%/100\% \text{ ME}} = 98.88 \text{ mhp, round to } \textbf{99 mhp}$$

114. **Find the whp for the following system: Motor efficiency is 90.5%; pump efficiency is 76%; and mhp is 220.**

Equation: **Water horsepower = (mhp)(ME)(PE)**

Water horsepower = (220 mhp)(90.5%/100% ME)(76%/100 PE)

Water horsepower = 151.316 whp, round to **150 whp**

115. **Find the whp, if the bhp is 41 and the pump efficiency is 81%.**

Equation: **Water hp = (bhp)(PE)**

Water hp = (41 bhp)(81%/100% PE) = 33.21 whp, round to **33 whp**

116. **What is the motor horsepower (mhp), if 110 horsepower (hp) is required to run a pump with a motor efficiency (ME) of 88% and a pump efficiency (PE) of 76%? *Note:* The 110 hp in this problem is called the water horsepower (whp). The whp is the actual energy (horsepower) available to pump water.**

Equation: **Motor horsepower** $= \dfrac{\text{whp}}{(\text{ME})(\text{PE})}$

$$\text{mhp} = \frac{110 \text{ whp}}{(88\%/100\% \text{ ME})(76\%/100\% \text{ PE})}$$

$$\text{mhp} = \frac{110 \text{ whp}}{(0.88 \text{ ME})(0.76 \text{ PE})}$$

mhp $= 164.47$ mhp, round to **160 mhp**

117. **Calculate the cost to run a pump in dollars and cents per day, if the horsepower is 88.5 and the cost per kW-hr is $0.079.**

Equation: **Cost, $/day = (mhp)(24 hr/day)(0.746 kW/hp)(Cost/kW-hr)**

Cost, $/day = (88.5 hp)(24 hr/day)(0.746 kW/hp)(0.079/kW-hr) = **$125.18/day**

DOSAGE PROBLEMS

These calculations are used mainly for process control, which requires accurate determination before the chemical is actually applied to a particular process. By keeping accurate records of dosages and thus usage, operators can also plan ordering or costing.

118. **How many lb/day of chlorine gas are required to treat 1,825,000 gpd given the following data?**

Chlorine demand = 8.5 mg/L
Chlorine residual = 1.25 mg/L

First, determine the total chlorine dose in mg/L.

Equation: **Chlorine dose = Chlorine demand + Chlorine residual**

Chlorine dose = 8.5 mg/L + 1.25 mg/L = 9.75 mg/L

Next, convert gpd to mgd.

Number of mgd = (1,825,000 gpd) / (1,000,000/mil) = 1.825 mgd

Lastly, calculate the lb/day of chlorine required.

Equation: **Number of lb/day = (Cl_2, mg/L)(Number of mgd)(8.34 lb/gal)**

Substitute values and solve.

Number of lb/day Cl_2 = (9.75 mg/L)(1.825 mgd)(8.34 lb/gal)

Number of lb/day Cl_2 = 148.4 lb/day, round to **150 lb/day Cl_2**

119. **What is the chlorine dosage in mg/L for a wastewater plant, if the chlorinator setting is 93 lb/day and it treats 2,100,000 gpd?**

First, convert gpd to mgd.

Number of mgd = (2,100,000 gpd) / (1,000,000/mil) = 2.1 mgd

Equation: **Number of lb/day = (Dose, mg/L)(Number of mgd)(8.34 lb/gal)**

Rearrange the equation to solve for the chlorine dosage in mg/L.

Chlorine dosage, mg/L $= \dfrac{\text{Number of lb/day}}{(\text{Number of mgd})\,(8.34\,\text{lb/gal})}$

Chlorine dosage, mg/L $= \dfrac{93\,\text{lb/day}}{(2.1\,\text{mgd})\,(8.34\,\text{lb/gal})} = 5.31$ mg/L, round to **5.3 mg/L**

120. **How many gpd of a 12.4% sodium hypochlorite solution are needed to disinfect a flow of 1,120,000 gallons, if the dosage required is 10.2 mg/L and the specific gravity of the hypochlorite is 1.04?**

First, convert gpd to mgd.

$$\text{Number of mgd} = \frac{1,120,000 \text{ gpd}}{1,000,000/\text{mil}} = 1.12 \text{ mgd}$$

Next, determine the lb/gal for the hypochlorite solution.

Hypochlorite, lb/gal = (8.34 lb/gal)(1.04 sp gr) = 8.6736

Next, using the "pounds equation," calculate the lb day of chlorine needed.

Equation: Chlorine, lb/day = (Dosage, mg/L)(mgd)(8.34 lb/gal)

Chlorine, lb/day = (10.2 mg/L)(1.12 mgd)(8.34 lb/gal) = 95.276 lb/day

Since the solution is not 100%, divide the percent hypochlorite into the lb/day of chlorine needed.

$$\text{Hypochlorite, lb/day} = \frac{95.276 \text{ lb/day}}{12.4\%/100\%} = 768.355 \text{ lb/day hypochlorite}$$

Lastly, determine the gpd of hypochlorite solution needed.

$$\text{Hypochlorite, gpd} = \frac{768.355 \text{ lb/day}}{8.6736 \text{ lb/gal}} = 88.58 \text{ gpd, round to } \mathbf{88.6 \text{ gpd Sodium hypochlorite}}$$

121. How much sulfur dioxide in lb/day needs to be applied to dechlorinate a waste-water, if the flow is 1,110,000 gpd, the average chlorine residual is 1.28 mg/L, the average chlorine demand is 5.88, and the sulfur dioxide must be 3.0 mg/L higher than the chlorine residual?

First, determine how many mg/L of sulfur dioxide (SO_2) must be applied. This is the chlorine residual plus the amount that is required higher than the chlorine residual. This additional amount of sulfur dioxide above the chlorine residual is applied as a safety factor and is typically started at 3 mg/L.

SO_2, mg/L = 1.28 mg/L + 5.88 mg/L + 3.0 mg/L = 10.16 mg/L SO_2

Next, convert gpd to mgd.

Number of mgd = (1,110,000 gpd) / (1,000,000/mil) = 1.11 mgd

Next, determine the number of lb/day of SO_2 needed.

Equation: **Number of lb/day SO_2 = (SO_2, mg/L)(Number of mgd)(8.34 lb/gal)**

Substitute values and solve.

Number of lb/day SO_2 = (10.16 mg/L)(1.11 mgd)(8.34 lb/gal)

Number of lb/day SO_2 = 94.05 lb/day, round to **94 lb/day SO_2**

122. **Determine the number of lb of sulfur dioxide (SO_2) used to treat a wastewater plant's effluent given the flowing data:**

Average flow per day = 2,450,000 gallons
Average chlorine residual = 1.08 mg/L
Average chlorine demand = 6.33 mg/L
SO_2 dosage 3.0 mg/L higher than chlorine residual

First, convert gallons per day to mgd.

Number of mgd = (2,450,000 gpd) / (1,000,000/mil) = 2.45 mgd

Next, determine the mg/L of SO_2 that must be applied.

Number of mg/L SO_2 = 1.08 mg/L + 6.33 mg/L + 3.0 mg/L = 10.41 mg/L SO_2

Lastly, determine the lb/day SO_2.

Equation: **Number of lb/day = (SO_2, mg/L)(Number of mgd)(8.34 lb/gal)**

Substitute values and solve.

Number of lb/day SO_2 = (10.41 mg/L)(2.45 mgd)(8.34 lb/gal) = **212.7 lb SO_2, round to 210 lb SO_2**

123. **What is the chlorine dosage in lb/day for a wastewater plant, if the dosage required is 7.8 mg/L, the flow is 489,000 gallons, and the calcium hypochlorite solution is 65% available chlorine?**

First, convert gallons per day to mgd.

Number of mgd = (489,000 gpd) / (1,000,000/mil) = 0.489 mgd

Equation: **Number of lb/day** $= \dfrac{\text{(mg/L)}\,\text{(Number of mgd)}\,\text{(8.34 lb/gal)}}{\text{Percent purity}/100\%}$

Substitute values and solve.

$$\text{Number of lb/day} = \frac{(7.8 \text{ mg/L})(0.489 \text{ mgd})(8.34 \text{ lb/gal})}{65\%/100\%}$$

Number of lb/day = 48.94 lb/day, round to **49 lb/day Calcium hypochlorite**

124. **A wastewater treatment plant is treating 875 gpm with a polymer solution that has a specific gravity (sp gr) of 1.22. If the chemical pump delivers 163 mL in exactly 5 minutes, what is the polymer dosage in mg/L?**

First, determine the number of mL/min the pump is feeding.

Polymer feed, mg/L = 163 mL/5 min = 32.6 mL/min

Next, determine the lb/gal for the polymer.

Polymer, lb/gal = (8.34 lb/gal)(1.22 sp gr) = 10.1748 lb/gal

Next, convert gpm to mgd.

$$\text{Number of mgd} = \frac{(875 \text{ gpm})(1,440 \text{ min/day})}{1,000,000/\text{mil}} = 1.26 \text{ mgd}$$

Now, calculate the dosage using the following equation:

$$\textbf{Polymer dosage, mg/L} = \frac{(\text{mL/min})(1,440 \text{ min/day})(\text{Polymer, lb/gal})}{(3,785 \text{ mL/gal})(\text{mgd})(8.34 \text{ lb/gal})}$$

Substitute values and solve.

$$\text{Polymer dosage, mg/L} = \frac{(32.6 \text{ mL/min})(1,440 \text{ min/day})(10.1748 \text{ lb/gal})}{(3,785 \text{ mL/gal})(1.26 \text{ mgd})(8.34 \text{ lb/gal})}$$

Polymer dosage, mg/L = **12.0 mg/L of Polymer**

125. **What is the number of lb/day of alum used by a wastewater plant, if the plant is treating 1,850 gpm with an alum dose of 13.5 mg/L? The alum is 49.2% aluminum sulfate.**

First, convert gpm to mgd.

$$\text{Number of mgd} = \frac{(1,850 \text{ gpm})(1,440 \text{ min/day})}{1,000,000/\text{mil}} = 2.664 \text{ mgd}$$

Next, calculate the number of lb/day of alum required.

$$\text{Equation: } \textbf{lb/day} = \frac{(\text{Alum dose, mg/L})(\text{mgd})(8.34 \text{ lb/gal})}{\text{Percent alum purity}/100\%}$$

$$\text{Alum, lb/day} = \frac{(13.5 \text{ mg/L})(2.664 \text{ mgd})(8.34 \text{ lb/gal})}{49.2\%/100\%} = \textbf{610 lb/day of Alum}$$

CHEMICAL FEED SOLUTION SETTINGS

As above, these calculations are used mainly for process control, which requires accurate determination before the chemical is actually applied to a particular process. Also as above, by keeping accurate records of dosages and thus usage, operators can plan ordering or costing.

126. **What should the chemical feeder be set on in mL/min, if the desired polymer dosage is 136 gpd?**

$$\text{Equation: } \textbf{Number of mL/min} = \frac{(\text{Number of gallons used})(3,785 \text{ mL/gal})}{1,440 \text{ min/day}}$$

Substitute values and solve.

$$\text{Polymer, mL/min} = \frac{(136 \text{ gal})(3,785 \text{ mL/gal})}{1,440 \text{ min/day}} = 357.47 \text{ mL/min, round to } \textbf{357 mL/min}$$

127. **What should the chemical feeder be set on in mL/min, if the desired polymer dosage is 81.8 gpd?**

$$\text{Equation: } \textbf{Number of mL/min} = \frac{(\text{Number of gallons used})(3,785 \text{ mL/gal})}{1,440 \text{ min/day}}$$

Substitute values and solve.

$$\text{Polymer, mL/min} = \frac{(81.8 \text{ gal})(3{,}785 \text{ mL/gal})}{1{,}440 \text{ min/day}} = \textbf{215 mL/min}$$

128. Given the following data, calculate the feed rate of a polymer solution in mL/min:

Influent flow = 1,330 gpm
Polymer dose = 7.35 mg/L
Polymer solution specific gravity = 1.35
Polymer percent purity = 39.7%

First, convert gpm to mgd.

$$\text{Number of mgd} = \frac{(1{,}330 \text{ gpm})(1{,}440 \text{ min/day})}{1{,}000{,}000/\text{mil}} = 1.9152 \text{ mgd}$$

Next, determine the lb/gal for the polymer.

$$\text{Polymer, lb/gal} = (8.34 \text{ lb/gal})(1.35 \text{ sp gr}) = 11.259 \text{ lb/gal}$$

Now, calculate the dosage using the following equation:

$$\textbf{Polymer dosage, mg/L} = \frac{(\text{mL/min})(1{,}440 \text{ min/day})(\text{Polymer, lb/gal})}{(3{,}785 \text{ mL/gal})(\text{mgd})(8.34 \text{ lb/gal})(\text{Percent Polymer})}$$

Rearrange the equation to solve for mL/min.

Polymer feed, mL/min =

$$\frac{(\text{Polymer dosage, mg/L})(3{,}785 \text{ mL/gal})(\text{mgd})(8.34 \text{ lb/gal})(\text{Percent Polymer})}{(1{,}440 \text{ min/day})(\text{Polymer, lb/gal})}$$

Substitute values and solve.

$$\text{Polymer feed mL/min} = \frac{(7.35 \text{ mg/L})(3{,}785 \text{ mL/gal})(1.9152 \text{ mgd})(8.34 \text{ lb/gal})(39.7\%/100\%)}{(1{,}440 \text{ min/day})(11.259 \text{ lb/gal})}$$

Polymer feed mL/min = 10.88 mL/min, round to **10.9 mL/min Polymer**

DRY CHEMICAL FEED SETTINGS

As with liquid dosing, accuracy in dosing dry chemicals is important, too. The more accurate the dosage calculation is, the more probability there will be for an operator to control a treatment process and the better the records for future referral.

129. **What is the feed rate of a dry chemical in lb/day, if a sample collection bowl collected 248.5 grams (g) in 30.0 minutes?**

Know: 454 grams = 1 pound

First, determine the number of g/min.

Number of grams = 248.5 g/30.0 min = 8.28 g/min

Equation: **Chemical, lb/day** $= \dfrac{(\text{Number of g/min})\,(1{,}440 \text{ min/day})}{454 \text{ g/lb}}$

Substitute values and solve.

Chemical, lb/day $= \dfrac{(8.28 \text{ g/min})\,(1{,}440 \text{ min/day})}{454 \text{ g/lb}} =$ **26.3 lb/day**

130. **What must have been the setting of a dry chemical feeder in grams/min, if the number of lb/day was 107.4?**

Equation: **Chemical, lb/day** $= \dfrac{(\text{Number of g/min})\,(1{,}440 \text{ min/day})}{454 \text{ g/lb}}$

Rearrange to solve for the feeder setting in g/min.

Number of g/min $= \dfrac{(\text{Chemical, lb/day})\,(454 \text{ g/lb})}{1{,}440 \text{ min/day}}$

Substitute values and solve.

Number of g/min $= \dfrac{(107.4 \text{ lb/day})\,(454 \text{ g/lb})}{1{,}440 \text{ min/day}} = 33.86$ g/min, round to **33.9 g/min**

131. Determine the feed rate of dry alum in lb/day, if the drawdown in exactly 10 minutes was 106.8 grams (g) and the flow is 1,090,000 gpd.

First, determine the number of grams used per minute.

Alum, g = 106.8 g/10 min = 10.68 g/min

Equation:

$$\textbf{Alum, lb/day} = \frac{(\text{Number of g/min})(1,440 \text{ min/day})}{454 \text{ g/lb}}$$

Substitute values and solve.

$$\text{Alum, lb/day} = \frac{(10.68 \text{ g/min})(1,440 \text{ min/day})}{454 \text{ g/lb}} = 33.87 \text{ lb/day, round to } \textbf{33.9 lb/day Alum}$$

132. What is the feed rate of a dry chemical in g/min, if the feed rate is 178 lb/day?

Equation: $\textbf{Chemical, lb/day} = \dfrac{(\text{Number of g/min})(1,440 \text{ min/day})}{454 \text{ g/lb}}$

Rearrange the equation to solve for g/min.

$$\text{Number of g/min} = \frac{(\text{Chemical, lb/day})(454 \text{ g/lb})}{1,440 \text{ min/day}}$$

Substitute values and solve.

$$\text{Number of g/min} = \frac{(178 \text{ lb/day})(454 \text{ g/lb})}{1,440 \text{ min/day}} = 56.12 \text{ g/min, round to } \textbf{56.1 g/min}$$

SLUDGE PRODUCTION CALCULATIONS

Sludge production calculations are important for costing and disposal purposes. Plants that use processes like digestion or heat treatment have smaller sludge production because more of the sludge is destroyed compared to plants that use chemical addition to treat wastes.

133. **A wastewater plant with an influent flow of 2,610,000 gpd has primary influent suspended solids of 187 mg/L. If the secondary suspended solids are 94 mg/L, what is the amount of dry solids produced in lb/day?**

First, determine the number of mg/L of suspended solids (SS) removed.

SS removed, mg/L = 187 mg/L, influent − 94 mg/L, effluent = 93 mg/L SS removed

Next, convert gpd to mgd.

$$\text{Number of mgd} = \frac{2,610,000 \text{ gpd}}{1,000,000/\text{mil}} = 2.61 \text{ mgd}$$

SS removed, lb/day = (SS removed, mg/L)(Number of mgd)(8.34 lb/gal)

SS removed, lb/day = (93 mg/L SS)(2.61 mgd)(8.34 lb/gal)

SS removed, lb/day = 2,024.37 lb/day, round to **2,000 lb/day**

134. **Given the following data, determine the amount of dry solids produced in lb/day:**

Flow = 3,080,000 gpd
Influent suspended solids = 268 mg/L
Primary effluent suspended solids = 98 mg/L

First, determine the number of mg/L of suspended solids (SS) removed.

SS removed, mg/L = 268 mg/L, influent − 98 mg/L effluent = 170 mg/L SS removed

Next, convert gpd to mgd.

Number of mgd $= \dfrac{3,080,000 \text{ gpd}}{1,000,000/\text{mil}} = 3.08$ mgd

SS removed, lb/day = (SS removed, mg/L)(Number of mgd)(8.34 lb/gal)

SS removed, lb/day = (170 mg/L SS)(3.08 mgd)(8.34 lb/gal)

SS removed, lb/day = 4,366.82 lb/day, round to **4,400 lb/day**

135. **Given the following data, determine the amount of flow the wastewater plant is treating in gpm.**

Primary effluent suspended solids (SS)= 141 mg/L
Primary effluent SS removed = 2,918 lb/day

Equation: **SS removed, lb/day = (SS removed, mg/L)(Number of mgd)(8.34 lb/gal)**

Rearrange to solve for mg/L SS removed.

Number of mgd $= \dfrac{\text{SS, lb/day removed}}{(\text{SS removed, mg/L})(8.34 \text{ lb/gal})}$

Number of mgd $= \dfrac{2,918 \text{ lb/day removed}}{(141 \text{ mg/L SS})(8.34 \text{ lb/gal})} = 2.48$ mgd

Next, convert mgd to gpm.

Number of gpm $= \dfrac{(2.48 \text{ mgd})(1,000,000/\text{mil})}{(1,440 \text{ min/day})} = 1,722$ gpm, round to **1,720 gpm**

136. **Given the following data, determine the amount of dry solids produce in lb/day:**

Flow = 3,180,000 gpd
Influent suspended solids (SS) = 385 mg/L
Primary effluent SS = 154 mg/L
Specific gravity (sp gr) of SS = 1.02

First, determine the number of mg/L of suspended solids removed.

SS removed, mg/L = 385 mg/L, influent − 154 mg/L, effluent = 231 mg/L SS removed

Next, convert the gpd to mgd.

$$\text{Number of gpd} = \frac{3,180,000 \text{ gpd}}{1,000,000/\text{mil}} = 3.18 \text{ mgd}$$

SS removed, lb/day = (SS removed, mg/L)(Number of mgd)(8.34 lb/gal)(SS, sp gr)

SS removed, lb/day = (231 mg/L SS)(3.18 mgd)(8.34 lb/gal)(1.02 sp gr)

SS removed, lb/day = 6,248.925 lb/day, round to **6,250 lb/day**

137. **A wastewater plant with an influent flow of 2,680,000 gpd has primary influent suspended solids of 372 mg/L. If the secondary suspended solids are 155 mg/L, and the specific gravity of the removed suspended solids is 1.01, what is the amount of dry solids produced in lb/day?**

First, determine the number of mg/L of suspended solids (SS) removed.

SS removed, mg/L = 372 mg/L, influent − 155 mg/L, effluent = 217 mg/L SS removed

Next, convert the gpd to mgd.

$$\text{Number of gpd} = \frac{2,680,000 \text{ gpd}}{1,000,000/\text{mil}} = 2.68 \text{ mgd}$$

SS removed, lb/day = (mg/L SS removed)(Number of mgd)(8.34 lb/gal)(SS, sp gr)

SS removed, lb/day = (217 mg/L SS)(2.68 mgd)(8.34 lb/gal)(1.01 sp gr)

SS removed, lb/day = 4,898.71 lb/day, round to **4,900 lb/day**

SLUDGE AGE (GOULD) CALCULATIONS

Operators need to understand sludge age calculations because this will help them maintain an appropriate amount of activated sludge in an aeration tank. The age of the sludge refers to the average solids retention time (usually in days) that the solids remain in the aeration tank. The sludge age is controlled by the sludge wasting rate, which affects the sludge yield in the system. This calculation is similar to detention time. See Figure 2 in Appendix E for one type of wastewater plant using an aeration tank.

138. Determine the age of the sludge in an aeration tank given the following data:

Primary effluent flow = 2.13 mgd
Volume of aeration tank = 495,000 gallons
Mixed liquor suspended solids (MLSS) = 1,930 mg/L
Suspended solids = 68 mg/L

First, convert the volume of the aeration tank to mil gal.

$$\text{Number of mil gal} = \frac{495,000 \text{ gal}}{1,000,000/\text{mil}} = 0.495 \text{ mil gal}$$

Equation: **Sludge age, days** $= \dfrac{(\text{MLSS, mg/L})(\text{Volume of aeration tank})(8.34 \text{ lb/gal})}{(\text{SS, mg/L})(\text{Flow, mgd})(8.34 \text{ lb/gal})}$

Sludge age, days $= \dfrac{(1,930 \text{ mg/L})(0.495 \text{ mil gal})(8.34 \text{ lb/gal})}{(68 \text{ mg/L})(2.13 \text{ mgd})(8.34 \text{ lb/gal})} =$ **6.6 days**

139. **Calculate the number of pounds of mixed liquor suspended solids (MLSS) that an aeration tank must maintain in order to keep the desired sludge age at 7.0 days. The amount of suspended solids in the primary effluent entering the aeration tank is 1,445 mg/L.**

Equation: **Number of lb MLSS = (Sludge age, days)(Suspended solids, lb/day)**

Substitute values and solve.

Number of lb MLSS = (7.0 days)(1,445 mg/L SS) = 10,115 lb MLSS, round to **10,000 lb MLSS**

SLUDGE VOLUME INDEX AND SLUDGE VOLUME DENSITY CALCULATIONS

The sludge volume index (SVI) and density calculations inform the operator about the flocculation process, the settling characteristics of the activated sludge, and how these settling characteristics affect return sludge rates and mixed liquor suspended solids. Activated sludge plants that are functioning properly usually have an SVI of around 100 mL/g.

140. **Given the following data, determine the sludge volume index (SVI):**

MLSS = 2,850 mg/L
SS = 688 mL
Sample size = 2,000 mL ± 1 mL

First, divide the SS result by 2 since the sample size is 2 liters and we need the SS value per liter.

SS, 1 liter = 688 mL/2 = 344 mL

Next, convert MLSS in mg to grams.

MLSS, g = (2,850 mg)(1 g/1,000 mg) = 2.85 g

Equation: $\text{SVI} = \dfrac{(\text{SS, mL})}{\text{MLSS, g/L}}$

Substitute values and solve.

$\text{SVI} = \dfrac{(344 \text{ mL})}{2.85 \text{ g}} = 120.7$, round to **121 SVI**

141. **A 30-minute settleability test had a settleable solids result of 241 mL in a 1-liter graduated cylinder. If the mixed liquor suspended solids (MLSS) in the aeration tank were 2,945 mg/L, what was the sludge density index (SDI)?**

First, convert MLSS in mg to grams.

MLSS, g = (2,945 mg)(1 g/1,000 mg) = 2.945 g

Equation: $\mathbf{SDI} = \dfrac{(MLSS, g)(100\%)}{SS, mL}$

Substitute values and solve.

$SDI = \dfrac{(2.945\ g)(100\%)}{241\ mL} = \mathbf{1.22\ SDI}$

ORGANIC LOADING RATE CALCULATIONS

Organic loading rate calculations tell the operator the amount of food entering the plant. These calculations are used for wastewater treatment ponds, rotating biological contactors, or trickling filters. See figures in Appendix E for the types of wastewater plants using these processes.

142. **What is the organic loading rate for a trickling filter that is 75.5 ft in diameter and 5.2 ft deep in lb BOD_5/d/1,000 ft³, if the primary effluent flow is 3.46 mgd and the BOD_5 is 168 mg/L?**

First, determine the volume of the tricking filter in ft³.

Volume, ft³ = (0.785)(Diameter)²(Depth, ft)

Volume, ft³ = (0.785)(75.5 ft)(75.5 ft)(5.2 ft) = 23,268.42 ft³

Next, factor out 1,000 ft³ from the volume = (23.26842)(1,000 ft³)

Next, determine the pounds of BOD_5/d/1,000 ft³ using a modified version of the "pounds" equation.

Organic loading rate, lb BOD$_5$/d/1,000 ft^3 $= \dfrac{(BOD_5,\ mg/L)\ (Flow,\ mgd)\ (8.34\ lb/gal)}{Volume\ of\ trickling\ filter,\ ft^3/1,000\ ft^3}$

Organic loading rate, lb BOD$_5$/d/1,000 ft^3 $= \dfrac{(168\ mg/L\ BOD_5)\ (3.46\ mgd)\ (8.34\ lb/gal)}{(23.26842)\ (1,000\ ft^3)}$

Organic loading rate, lb BOD$_5$/d/1,000 ft^3 = 208.346 lb BOD$_5$/d/1,000 ft, round to 210 lb BOD$_5$/d/1,000 ft^3

143. **Given the following data on a wastewater treatment pond, calculate the organic loading rate in lb BOD$_5$/d/acre-ft:**

Influent flow = 278,000 gpd
Surface area of pond = 3.49 acre-ft
Influent BOD$_5$ concentration = 287 mg/L

First, convert gallons per day to mgd.

Number of mgd $= \dfrac{278,000\ gpd}{1,000,000/mil} = 0.278\ mgd$

Next, determine the pounds of BOD$_5$/d/acre-ft using a modified version of the "pounds" equation.

Organic loading rate, lb BOD$_5$/d/acre-ft $= \dfrac{(BOD_5,\ mg/L)\ (Flow,\ mgd)\ (8.34\ lb/gal)}{Surface\ area\ of\ pond,\ acre-ft}$

Organic loading rate, lb BOD$_5$/d/acre-ft $= \dfrac{(287\ mg/L\ BOD_5)\ (0.278\ mgd)\ (8.34\ lb/gal)}{3.49\ acre-ft}$

Organic loading rate, lb BOD$_5$/d/acre-ft = 190.66 lb BOD$_5$/d/acre-ft, round to
191 lb BOD$_5$/d/acre-ft

144. **What is the organic loading rate for a trickling filter in pounds biochemical oxygen demand per day per 1,000 cubic feet (lb BOD$_5$/d/1,000 ft^3) given the following data?**

Trickling filter radius = 37.6 ft
Trickling filter depth = 6.3 ft
BOD$_5$ = 173 mg/L
Primary effluent flow = 3,275,000 gpd

First, convert gallons per day to mgd.

$$\text{Number of mgd} = \frac{3,275,000 \text{ gpd}}{1,000,000/\text{mil}} = 3.275 \text{ mgd}$$

Next, calculate the volume of the trickling filter.

Know: **Volume, ft^3 = πr^2(Depth)** where r equals the radius

Volume, ft^3 = (3.14)(37.6 ft)(37.6 ft)(6.3 ft) = 27,967 ft^3

Next, factor out 1,000 ft^3 from the volume = (27.967)(1,000 ft^3)

Organic loading rate, lb BOD$_5$/d/1,000 ft^3 $= \dfrac{(\text{BOD}_5, \text{mg/L})\,(\text{Flow, mgd})\,(8.34 \text{ lb/gal})}{(\text{Volume of trickling filter})\,(1,000 \text{ ft}^3)}$

Organic loading rate, lb BOD$_5$/d/1,000 ft^3 $= \dfrac{(173 \text{ mg/L BOD}_5)\,(3.275 \text{ mgd})\,(8.34 \text{ lb/gal})}{(27.967)\,(1,000 \text{ ft}^3)}$

Organic loading rate, lb BOD$_5$/d/1,000 ft^3 = 168.96 lb BOD$_5$/d/1,000 ft^3, round to 170 lb BOD$_5$/d/1,000 ft^3

145. **What is the organic loading rate for a rotating biological contactor (RBC) in lb BOD$_5$/d/1,000 ft^2 given the following data?**

Surface area of RBC = 760,000 ft^2
BOD$_5$ = 109 mg/L
Flow = 4,805,000 gpd

First, convert gallons per day to mgd.

$$\text{Number of mgd} = \frac{4,805,000 \text{ gpd}}{1,000,000/\text{mil}} = 4.805 \text{ mgd}$$

Next, factor out 1,000 ft^2 from the surface area = (760)(1,000 ft^2)

Organic loading rate, lb BOD$_5$/d/1,000 ft^2 $= \dfrac{(\text{BOD}_5, \text{mg/L}) (\text{Flow, mgd}) (8.34 \text{ lb/gal})}{(\text{Surface area of RBC}) (1,000 \text{ ft}^2)}$

Organic loading rate, lb BOD$_5$/d/1,000 ft^2 $= \dfrac{(109 \text{ mg/L BOD}_5) (4.805 \text{ mgd}) (8.34 \text{ lb/gal})}{(760) (1,000 \text{ ft}^2)}$

Organic loading rate, lb BOD$_5$/d/1,000 ft^2 = 5.75 lb BOD$_5$/d/1,000 ft^2, round to 5.8 lb BOD$_5$/d/1,000 ft^2

SUSPENDED SOLIDS LOADING CALCULATIONS

Operators use suspended solids loading for evaluating process control.

146. **Calculate the amount of suspended solids entering a trickling filter in lb/day, if the influent flow is 875 gallons per minute (gpm) and the amount of suspended solids (SS) is 255 mg/L?**

First, determine the number of mgd.

$$\text{Number of mgd} = \frac{(875 \text{ gpm}) (1,440 \text{ min/day})}{1,000,000/\text{mil}} = 1.26 \text{ mgd}$$

This problem uses the same equation as a dosage problem.

Equation: **Number of lb/day SS = (SS, mg/L)(Number of mgd)(8.34 lb/gal)**

Substitute values and solve.

Number of lb/day SS = (255 mg/L SS)(1.26 mgd)(8.34 lb/gal)

Number of lb/day SS = 2,679.64 lb/day, round to **2,680 lb/day SS**

147. **If the influent flow to a trickling filter is 995 gpm and the suspended solids load-ing is 2,850 lb/day, what is the suspended solids (SS) entering a trickling filter in mg/L?**

First, determine the number of mgd.

$$\text{Number of mgd} = \frac{(995 \text{ gpm})(1,440 \text{ min/day})}{1,000,000/\text{mil}} = 1.4328 \text{ mgd}$$

Equation: **Number of lb/day = (SS, mg/L)(Number of mgd)(8.34 lb/gal)**

Rearrange the equation to solve for mg/L, SS.

$$\text{Number of mg/L SS} = \frac{\text{Number of lb/day}}{(\text{Number of mgd})(8.34 \text{ lb/gal})}$$

Substitute values and solve.

$$\text{Number of mg/L SS} = \frac{2,850 \text{ lb/day}}{(1.4328 \text{ mgd})(8.34 \text{ lb/gal})} = 238.5 \text{ mg/L, round to } \textbf{238 mg/L SS}$$

148. **What is the amount of suspended solids entering a trickling filter in lb/day, if the influent flow is 650 gpm and the amount of suspended solids (SS) is 343 mg/L?**

First, convert gallons per day to mgd.

$$\text{Number of mgd} = \frac{(650 \text{ gpm})(1,440 \text{ min/day})}{1,000,000/\text{mil}} = 0.936 \text{ mgd}$$

Then, calculate the lb/day of suspended solids.

Equation: **Number of lb/day = (SS, mg/L)(Number of mgd)(8.34 lb/gal)**

Substitute values and solve.

Number of lb/day SS = (343 mg/L SS)(0.936 mgd)(8.34 lb/gal)

Number of lb/day SS = 2,677.54 lb/day, round to **2,700 lb/day SS**

149. **If the influent flow entering a trickling filter is 1,350 gpm and the suspended solids loading is 3,888 lb/day, calculate the amount of suspended solids (SS) entering a trickling filter in mg/L.**

First, convert gallons per day to mgd.

$$\text{Number of mgd} = \frac{(1,350 \text{ gpm})(1,440 \text{ min/day})}{1,000,000/\text{mil}} = 1.944 \text{ mgd}$$

Equation: **Number of lb/day = (SS, mg/L)(Number of mgd)(8.34 lb/gal)**

Rearrange the equation to solve for mg/L, SS.

$$\text{Number of mg/L SS} = \frac{\text{Number of lb/day}}{(\text{Number of mgd})(8.34 \text{ lb/gal})}$$

Substitute values and solve.

$$\text{Number of mg/L SS} = \frac{3,888 \text{ lb/day}}{(1.944 \text{ mgd})(8.34 \text{ lb/gal})} = 239.8 \text{ mg/L, round to } \textbf{240 mg/L SS}$$

BIOCHEMICAL OXYGEN DEMAND LOADING CALCULATIONS

Biochemical oxygen demand is the demand for oxygen made by bacteria as they decompose organic matter in wastewater or in the natural environment. This calculation sometimes is helpful in evaluating treatment pond processes. The BOD_5 is a 5-day test. See Figures 1 and 6 in Appendix E for two types of wastewater plants using a trickling filter.

150. **What is the biochemical oxygen demand (BOD_5) loading on a trickling filter in lb/day, if the influent flow into the trickling filter is 1,250 gpm and the influent BOD_5 is 263 mg/L?**

First, determine the number of mgd.

$$\text{Number of mgd} = \frac{(1,250 \text{ gpm})(1,440 \text{ min/day})}{1,000,000/\text{mil}} = 1.80 \text{ mgd}$$

Equation: **Number of lb/day BOD_5 = (BOD_5, mg/L)(Number of mgd)(8.34 lb/gal)**

Substitute values and solve.

Number of lb/day BOD_5 = (263 mg/L BOD_5)(1.80 mgd)(8.34 lb/gal)

Number of lb/day BOD_5 = 3,948.156 lb/day BOD_5, round to **3,950 lb/day BOD_5**

151. **What is the influent flow to a trickling filter in gpm, if the BOD$_5$ loading is 3,045 lb/day and the BOD$_5$ is 279 mg/L?**

Equation: **Number of lb/day BOD$_5$ = (BOD$_5$, mg/L)(Number of mgd)(8.34 lb/gal)**

Rearrange the equation to solve for mgd.

$$\textbf{Number of mgd} = \frac{\text{Number of lb/day BOD}_5}{(\text{Number of mg/L BOD}_5)(8.34\ \text{lb/gal})}$$

Substitute values and solve.

$$\text{Number of mgd flow} = \frac{3,045\ \text{lb/day}}{(279\ \text{mg/L})(8.34\ \text{lb/gal})} = 1.3086\ \text{mgd}$$

Lastly, convert mgd to gpm.

$$\text{Number of gpm} = \frac{(1.3086\ \text{mgd})(1,000,000/\text{mil})}{1,440\ \text{min/day}} = \textbf{909 gpm}$$

152. **Given the following data, calculate the biochemical oxygen demand (BOD$_5$) loading on a trickling filter in lb/day:**

Influent flow = 3,440,000 gallons per day (gpd)
Influent BOD$_5$ = 361 mg/L
Effluent BOD$_5$ = 99 mg/L

First, convert gallons to mgd.

$$\text{Number of mgd} = \frac{3,440,000\ \text{gpd}}{1,000,000/\text{mil}} = 3.44\ \text{mgd}$$

Then, calculate the BOD$_5$ loading in lb/day.

Equation: **Number of lb/day BOD$_5$ = (BOD$_5$, mg/L)(Number of mgd)(8.34 lb/gal)**

Substitute values and solve.

Number of lb/day BOD$_5$ = (361 mg/L BOD$_5$)(3.44 mgd)(8.34 lb/gal)

Number of lb/day BOD$_5$ = 10,356.946 lb/day, round to **10,000 lb/day BOD$_5$**

153. Calculate the influent flow to a trickling filter in gpm, if the BOD_5 loading is 3,885 lb/day and the BOD_5 is 371 mg/L.

First, solve for number of mgd by using the "pounds" equation.

Equation: **Number of lb/day BOD_5 = (BOD_5, mg/L)(Number of mgd)(8.34 lb/gal)**

Rearrange the equation to solve for mgd.

$$\textbf{Number of mgd} = \frac{\text{Number of lb/day BOD}_5}{(\text{Number of mg/L BOD}_5)(8.34\,\text{lb/gal})}$$

Substitute values and solve.

$$\text{Number of mgd flow} = \frac{3,885\,\text{lb/day}}{(371\,\text{mg/L})(8.34\,\text{lb/gal})} = 1.256\,\text{mgd}$$

Then, convert mgd to gpm.

$$\text{Number of gpm} = \frac{(1.256\,\text{mgd})(1,000,000/\text{mil})}{1,440\,\text{min/day}} = 872.22\,\text{gpm, round to } \textbf{872 gpm}$$

Another way to solve this problem is to write all the values at once.

$$\text{Number of gpm} = \frac{(3,885\,\text{lb/day})(1,000,000/\text{mil})}{(371\,\text{mg/L})(8.34\,\text{lb/gal})(1,440\,\text{min/day})} = 871.94\,\text{gpm, round to } \textbf{872 gpm}$$

Note: The multiple-step approach may be easier for novices, but the later method is faster. The key is always to make sure all units cancel except those sought after.

SOLUBLE AND PARTICULATE BIOCHEMICAL OXYGEN DEMAND CALCULATIONS

BOD_5 measures the amount of organic matter that is present in water. Bacteria break down this organic matter by natural decomposition and in the process utilize oxygen. Thus the more organic matter present in the wastewater, the more demand for oxygen by the bacteria. Operators need to know how to do BOD_5 calculations because there are strict regulations for the amount of BOD_5 that can be discharged to a natural water body from the treated plant. The K value in these problems is the portion of suspended solids in the wastewater that are organic suspended solids. Domestic water usually has about 50 to 70% of the suspended solids as organic suspended solids, which is usually written in decimal form (0.5 to 0.7).

154. **Given the following, calculate the estimated particulate BOD_5 in mg/L for the following wastewater.**

Plant Flow = 1,775 gpm
SS solids = 265 mg/L
K factor = 0.63

Equation: **Particulate BOD_5, mg/L = (SS, mg/L)(K factor)**

Substitute values and solve.

Particulate BOD_5, mg/L = (265 mg/L SS)(0.63 K value)

Particulate BOD_5, mg/L = 166.95 mg/L, round to **170 mg/L Particulate BOD_5**

155. **What is the soluble BOD_5, if the suspended solids are 153 mg/L, the total BOD_5 is 189 mg/L, and the K factor is 0.67?**

Equation: **Total BOD_5 = (Particulate BOD_5)(K factor) + Soluble BOD_5**

Rearrange the equation to solve for soluble BOD_5.

Soluble BOD_5 = Total BOD_5 − (Particulate BOD_5)(K factor)

Substitute values and solve.

Soluble BOD_5 = 189 mg/L BOD_5 − (153 mg/L BOD_5)(0.67 K factor)

Soluble BOD_5 = 189 mg/L BOD_5 − 102.51 mg/L BOD_5

Soluble BOD_5 = 86.49 mg/L, round to **86 mg/L Soluble BOD_5**

156. What is the particulate BOD_5, if the total BOD_5 is 189 mg/L, the soluble BOD_5 is 106 mg/L, and the K factor is 0.61?

Equation: **Total BOD_5 = (Particulate BOD_5)(K factor) + Soluble BOD_5**

Rearrange the equation to solve for particulate BOD_5, by first subtracting soluble BOD_5 from both sides of the equation.

(Particulate BOD_5)(K factor) = Total BOD_5 − Soluble BOD_5

Then, divide both sides by the K factor.

$$\text{Particulate } BOD_5 = \frac{\text{Total } BOD_5 - \text{Soluble } BOD_5}{\text{K factor}}$$

Substitute values and solve.

$$\text{Particulate } BOD_5 = \frac{189 \text{ mg/L } BOD_5 - 106 \text{ mg/L } BOD_5}{0.61 \text{ factor}}$$

Particulate BOD_5 = 136.07 mg/L, round to **140 mg/L Particulate BOD_5**

157. How many lb/day of soluble BOD_5 enters a rotating biological contactor (RBC) each day, if the flow is 1.02 mgd, total BOD_5 is 217 mg/L, the particulate BOD_5 is 131 mg/L, and the K factor is 0.59?

Equation: **Total BOD_5 = (Particulate BOD_5)(K factor) + Soluble BOD_5**

Rearrange the equation to solve for soluble BOD_5 by subtracting (Particulate BOD_5)(K factor) from each side.

Soluble BOD_5 = Total BOD_5 − (Particulate BOD_5)(K factor)

Substitute values and solve.

Soluble BOD_5 = 217 mg/L BOD_5 − (131 mg/L BOD_5)(0.59 K factor)

Soluble BOD_5 = 217 mg/L BOD_5 − 77.29 mg/L BOD_5

Soluble BOD_5 = 139.71 mg/L Soluble BOD_5

Next, determine the Soluble BOD_5 in lb/day.

Equation: **Soluble BOD_5, lb/day = (Soluble BOD_5, mg/L)(mgd)(8.34 lb/gal)**

Soluble BOD_5, lb/day = (139.71 mg/L)(1.02 mgd)(8.34 lb/gal)

Soluble BOD_5, lb/day = 1,188.48 lb/day BOD_5, round to **1,200 lb/day BOD_5**

CHEMICAL OXYGEN DEMAND LOADING CALCULATIONS

Chemical oxygen demand is a measure of the capacity of water to consume oxygen, when organic matter and the oxidation of inorganic matter are decomposed. Because it also measures the decomposition of inorganic matter such as nitrate and ammonia, it is only an indirect measure of the organic matter in water.

158. **Calculate the chemical oxygen demand (COD) in lb/day that is applied to an aeration tank, if the flow is 2,500,000 gallons and the COD concentration is 120 mg/L.**

First, convert the gpd to mgd.

$$\text{Number of gpd} = \frac{2,500,000 \text{ gpd}}{1,000,000/\text{mil}} = 2.5 \text{ mgd}$$

Equation: **Number of lb/day COD = (COD, mg/L)(Number of mgd)(8.34 lb/gal)**

Substitute values and solve.

Number of lb/day = (120 mg/L)(2.5 mgd)(8.34 lb/gal)

Number of lb/day = 2,502 lb/day, round to **2,500 lb/day**

159. **If the influent flow to an aeration tank is 1,390,000 gpd and the chemical oxygen demand (COD) loading is 2,565 lb/day, what is the amount of COD entering an aeration tank in mg/L?**

First, convert gpd to mgd.

$$\text{Number of mgd} = \frac{1,390,000 \text{ gpd}}{1,000,000/\text{mil}} = 1.39 \text{ mgd}$$

Next, using the following equation calculate the COD, mg/L.

Equation: **Number of lb/day COD = (COD, mg/L)(Number of mgd)(8.34 lb/gal)**

Rearrange the equation to solve for mg/L.

$$\text{Number of mg/L COD} = \frac{\text{Number of lb/day COD}}{(\text{Number of mgd})(8.34 \text{ lb/gal})}$$

Substitute values and solve.

$$\text{Number of mg/L COD} = \frac{2,565 \text{ lb/day}}{(1.39 \text{ mgd})(8.34 \text{ lb/gal})} = 221.26 \text{ mg/L, round to } \textbf{221 mg/L COD}$$

160. **If the COD loading to an aeration tank is 2,890 lb/day and the COD is 145 mg/L, what is the influent flow to the aeration tank in mgd?**

Equation: **Number of lb/day COD = (COD, mg/L)(Number of mgd)(8.34 lb/gal)**

Rearrange the equation to solve for mgd.

$$(\text{Number of mgd}) = \frac{\text{Number of lb/day COD}}{(\text{Number of mg/L COD})(8.34 \text{ lb/gal})}$$

Substitute values and solve.

$$\text{Number of mgd flow} = \frac{2,890 \text{ lb/day}}{(145 \text{ mg/L})(8.34 \text{ lb/gal})} = \textbf{2.39 mgd}$$

161. What is the chemical oxygen demand (COD) entering an aeration tank in mg/L, if the influent flow is 0.645 mgd and the COD loading is 1,020 lb/day?

Equation: **Number of lb/day COD = (COD, mg/L)(Number of mgd)(8.34 lb/gal)**

Rearrange the equation to solve for mg/L.

$$\text{Number of mg/L COD} = \frac{\text{Number of lb/day COD}}{(\text{Number of mgd})(8.34\,\text{lb/gal})}$$

Substitute values and solve.

$$\text{Number of mg/L COD} = \frac{1,020\,\text{lb/day}}{(0.645\,\text{mgd})(8.34\,\text{lb/gal})} = 189.62\,\text{mg/L, round to \textbf{190 mg/L COD}}$$

162. What is the influent flow to an aeration tank in gpm, if the COD loading is 2,515 lb/day and the COD is 211 mg/L?

First, determine the number of mgd.

Equation: **Number of lb/day COD = (COD, mg/L)(Number of mgd)(8.34 lb/gal)**

Rearrange the equation to solve for mgd.

$$\text{Number of mgd} = \frac{\text{Number of lb/day COD}}{(\text{Number of mg/L COD})(8.34\,\text{lb/gal})}$$

Substitute values and solve.

$$\text{Number of mgd flow} = \frac{2,515\,\text{lb/day}}{(211\,\text{mg/L})(8.34\,\text{lb/gal})} = 1.4292\,\text{mgd}$$

Then, convert mgd to gpm.

$$\text{Number of gpm} = \frac{(1.4292\,\text{mgd})(1,000,000/\text{mil})}{1,440\,\text{min/day}} = 992.5\,\text{gpm, round to \textbf{992 gpm}}$$

HYDRAULIC DIGESTION TIME CALCULATIONS

Hydraulic digestion time tells the operator how long the process will take to complete and is thus used for planning purposes. See Figures 2, 4, 5, and 6 in Appendix E for four types of wastewater plants using a digester.

163. **What is the hydraulic digestion time for a digester that is 24.5 ft in radius, has a level of 10.4 ft, and has sludge flow of 7,690 gallons per day (gpd)?**

First, determine the volume of the digester in gallons.

Volume, gal $= \pi r^2(\text{Depth, ft})(7.48 \text{ gal/ft}^3)$

Volume, gal $= (3.14)(24.5 \text{ ft})(24.5 \text{ ft})(10.4 \text{ ft})(7.48 \text{ gal/ft}^3) = 146{,}621.19 \text{ gal}$

Next, calculate the digestion time in days.

Equation: **Digestion time, days** $= \dfrac{\text{Number of gallons}}{\text{Influent sludge flow, gpd}}$

Substitute values and solve.

Digestion time, days $= \dfrac{146{,}621.19 \text{ gal}}{7{,}690 \text{ gpd}} = 19.07$ days, round to **19.1 days**

164. **What is the hydraulic digestion time for a 49.85-ft diameter digester with a level of 11.8 ft and sludge flow of 11,105 gallons per day (gpd)?**

Equation: **Digestion time, days** $= \dfrac{(0.785)(\text{Diameter})^2(\text{Depth, ft})(7.48 \text{ gal/ft}^3)}{\text{Influent sludge flow, gpd}}$

Substitute values and solve.

Digestion time, days $= \dfrac{(0.785)(49.85 \text{ ft})(49.85 \text{ ft})(11.8 \text{ ft})(7.48 \text{ gal/ft}^3)}{11{,}105 \text{ gpd}}$

Digestion time, days $=$ **15.5 days**

DIGESTER LOADING RATE CALCULATIONS

Digester loading rate calculations tell the operator how much volatile solids are stabilized per cubic foot of digester space. It is used for evaluating process control.

165. **What is the loading on a digester in lb volatile solids (VS)/day/ft³, if the sludge flow into the digester is 22,500 gpd, the digester is 50.5 ft in diameter, the sludge level is 21.5 ft, and the sludge is 5.15% solids with a specific gravity of 1.04 and 71.4% volatile solids?**

Equation: **Digester loading, lb VS/day/ft³ =**

$$\frac{(\text{Flow, gpd})(8.34\ \text{lb/gal})(\text{sp gr})(\text{Percent sludge})(\text{Percent volatile solids})}{(0.785)(\text{Diameter})^2(\text{Sludge level})}$$

Substitute values and solve.

$$\text{Digester loading, lb VS/day/ft}^3 = \frac{(22,500\ \text{gpd})(8.34\ \text{lb/gal})(1.04)(5.15\%/100\%)(71.4\%/100\%)}{(0.785)(50.5\ \text{ft})(50.5\ \text{ft})(21.5\ \text{ft})}$$

$$\text{Digester loading, lb VS/day/ft}^3 = \frac{7,176.08\ \text{lb VS/day}}{43,041.84\ \text{ft}^3}$$

$$\text{Digester loading, lb VS/day/ft}^3 = \mathbf{0.167\ lb\ VS/day/ft^3}$$

166. **Given the following data, calculate the digester loading in lb volatile solids (VS)/day/1,000 ft³.**

Digester diameter = 49.5 ft
Sludge level = 16.3 ft
Influent sludge flow = 17,450 gpd
Percent sludge solids = 4.95%
Percent volatile solids = 74.25%
Specific gravity of sludge = 1.05

Equation: **Digester loading, lb VS/day/1,000 ft³ =**

$$\frac{(\text{Flow, gpd})(8.34\ \text{lb/gal})(\text{sp gr})(\text{Percent sludge})(\text{Percent volatile solids})}{(0.785)(\text{Diameter})^2(\text{Sludge level})}$$

Substitute values and solve.

Digester loading, lb VS/day/1,000 ft^3 =

$$\frac{(17,450 \text{ gpd})(8.34 \text{ lb/gal})(1.05 \text{ sp gr})(4.95\%/100\%)(74.25\%/100\%)}{(0.785)(49.5 \text{ ft})(49.5 \text{ ft})(16.3 \text{ ft})}$$

Digester loading, lb VS/day/1,000 ft^3 = $\dfrac{5,616.33 \text{ lb VS/day}}{31,352.17 \text{ ft}^3}$

Factor out 1,000 ft^3 from the denominator and do not divide by 1,000 ft^3, as it will become part of the units and not part of the calculation.

Digester loading, lb VS/day/1,000 ft^3 = $\dfrac{5,616.33 \text{ lb VS/day}}{(31.35217)(1,000 \text{ ft}^3)}$

Digester loading, lb VS/day/1,000 ft^3 = $\dfrac{179.14 \text{ lb VS/day}}{1,000 \text{ ft}^3}$, round to **179 lb VS/day/1,000 ft^3**

167. Given the following data, calculate the digester loading in lb VS/day/1,000 ft^3.

Digester diameter = 50.2 ft
Sludge level = 18.7 ft
Influent sludge flow = 19,800 gpd
Percent sludge solids = 4.85%
Percent volatile solids = 70.5%
Specific gravity of sludge = 1.05

Equation: **Digester loading, lb VS/day/1,000 ft^3 =**

$$\frac{(\text{Flow, gpd})(8.34 \text{ lb/gal})(\text{sp gr})(\text{Percent sludge})(\text{Percent volatile solids})}{(0.785)(\text{Diameter})^2(\text{Sludge level})}$$

Substitute values and solve.

Digester loading, lb VS/day/1,000 ft^3 =

$$\frac{(19,800 \text{ gpd})(8.34 \text{ lb/gal})(1.05 \text{ sp gr})(4.85\%/100\%)(70.5\%/100\%)}{(0.785)(50.2 \text{ ft})(50.2 \text{ ft})(18.7 \text{ ft})}$$

Digester loading, lb VS/day = $\dfrac{5,928.59 \text{ lb VS/day}}{36,992.93 \text{ ft}^3}$

Factor out 1,000 ft³ from the denominator, as shown in the above problem.

$$\text{Digester loading, lb VS/day} = \frac{5{,}928.59 \text{ lb VS/day}}{(36.99293)(1{,}000 \text{ ft}^3)}$$

$$\text{Digester loading, lb VS/day/1,000 ft}^3 = \frac{160.26 \text{ lb VS/day}}{1{,}000 \text{ ft}^3}, \text{round to } \mathbf{160 \text{ lb VS/day/1,000 ft}^3}$$

MEAN CELL RESIDENCE TIME (SOLIDS RETENTION TIME) CALCULATIONS

The mean cell residence time (MCRT) is the average time the activated-sludge solids are in an activated biosolids system. The MCRT is an important design and operating parameter for operators to use in the activated-sludge process and is normally expressed in days. This calculation is used for operational process control. See Figure 2 in Appendix E for one type of wastewater plant using the activated sludge process.

168. What is the mean cell residence time (MCRT) for the following activated sludge system:

Aeration tank and final clarifier volume = 0.756 mil gal
Mixed liquor suspended solids (MLSS) = 3,180 mg/L
Suspended solids (SS) wasted = 1,750 lb/day
Secondary effluent SS = 416 lb/day

Equation: $\mathbf{MCRT, days} = \dfrac{(\text{MLSS, mg/L})(\text{mil gal})(8.34 \text{ lb/gal})}{\text{SS wasted, lb/day} + \text{SS lb/day}}$

Substitute values and solve.

$$\text{MCRT, days} = \frac{(3{,}180 \text{ mg/L})(0.756 \text{ mil gal})(8.34 \text{ lb/gal})}{1{,}750 \text{ lb/day SS} + 416 \text{ lb/day SS}}$$

$$\text{MCRT, days} = \frac{(3{,}180 \text{ mg/L})(0.756 \text{ mil gal})(8.34 \text{ lb/gal})}{2{,}166 \text{ lb/day SS}} = \mathbf{9.26 \text{ days}}$$

169. Given the following data, calculate the mean cell residence time (MCRT):

Flow = 2.72 mgd
Aeration tank volume = 603,000 gallons
Clarifier tank volume = 305,000 gallons
MLSS = 2,990 mg/L
Waste rate = 23,700 gpd
Waste activated sludge (WAS) = 7,175 mg/L
Effluent total suspended solids (TSS) = 15.0 mg/L

First, convert the volumes for the tanks to mil gal.

$$\text{Aeration tank, mgd} = \frac{603,000 \text{ gal}}{1,000,000/\text{mil}} = 0.603 \text{ mil gal}$$

$$\text{Clarifier tank, mgd} = \frac{305,000 \text{ gal}}{1,000,000/\text{mil}} = 0.305 \text{ mil gal}$$

Next, convert the waste rate from gpd to mgd.

$$\text{Waste rate, mgd} = \frac{23,700 \text{ gal}}{1,000,000/\text{mil}} = 0.0237 \text{ mgd}$$

Next, calculate the MCRT in days.

Equation: **MCRT, days =**

$$\frac{(\text{MLSS, mg/L})(\text{Aeration tank mil gal} + \text{Clarifier tank mil gal})(8.34 \text{ lb/gal})}{(\text{WAS, mg/L})(\text{Waste rate, mgd})(8.34 \text{ lb/gal}) + (\text{TSS, mg/L})(\text{Flow, mgd})(8.34 \text{ lb/gal})}$$

$$\text{MCRT, days} = \frac{(2,990 \text{ mg/L MLSS})(0.603 \text{ mil gal} + 0.305 \text{ mil gal})(8.34 \text{ lb/gal})}{(7,175 \text{ mg/L})(0.0237 \text{ mgd})(8.34 \text{ lb/gal}) + (15.0 \text{ mg/L TSS})(2.72 \text{ mgd})(8.34 \text{ lb/gal})}$$

$$\text{MCRT, days} = \frac{(2,990 \text{ mg/L MLSS})(0.908 \text{ mil gal})(8.34 \text{ lb/gal})}{1,418.196 \text{ lb/day} + 340.272 \text{ lb/day}}$$

$$\text{MCRT, days} = \frac{22,642.4328 \text{ lb}}{1,758.468 \text{ lb/day}} = 12.876 \text{ days, round to } \textbf{12.9 days}$$

170. **If the MCRT desired was 8.20 days, what would the waste rate be for the following system in lb/day?**

Flow = 1.34 mgd
Aeration tank (AT) volume = 0.485 mil gal
Clarifier tank (CT) volume = 0.309 mil gal
MLSS = 2,825 mg/L
Effluent TSS = 16.5 mg/L

Equation: **Waste rate, lb/day =**

$$\frac{\text{MLSS, mg/L}\,[\text{AT, mil gal} + \text{CT, mil gal}]\,(8.34\,\text{lb/gal})}{\text{Desired MCRT}} - (\text{TSS, mg/L})(\text{Flow, mgd})(8.34\,\text{lb/gal})$$

Waste rate, lb/day =

$$\frac{(2,825\,\text{mg/L})\,[0.485\,\text{mil gal} + 0.309\,\text{mil gal}]\,(8.34\,\text{lb/gal})}{8.20\,\text{days, Desired MCRT}} - (16.5\,\text{mg/L TSS})(1.34\,\text{mgd})(8.34\,\text{lb/gal})$$

Waste rate, lb/day =

$$\frac{(2,825\,\text{mg/L})\,(0.794\,\text{mil gal})\,(8.34\,\text{lb/gal})}{8.20\,\text{days, Desired MCRT}} - (16.5\,\text{mg/L TSS})(1.34\,\text{mgd})(8.34\,\text{lb/gal})$$

Waste rate, lb/day = 2,281.346 lb/day − 184.3974 lb/day = **2,100 lb/day**

171. **What is the mean cell residence time (MCRT), given the following data?**

Flow = 2.67 mgd
Aeration tank volume = 675,000 gallons
Clarifier tank volume = 322,000 gallons
MLSS = 2,840 mg/L
Waste rate = 25,700 gpd
Waste activated sludge (WAS) = 6,825 mg/L
Effluent TSS = 18.0 mg/L

First, convert the volumes for the tanks to mil gal.

$$\text{Aeration tank, mgd} = \frac{675,000 \text{ gal}}{1,000,000/\text{mil}} = 0.675 \text{ mil gal}$$

$$\text{Clarifier tank, mgd} = \frac{322,000 \text{ gal}}{1,000,000/\text{mil}} = 0.322 \text{ mil gal}$$

Next, convert the waste rate from gpd to mgd.

$$\text{Waste rate, mgd} = \frac{25,700 \text{ gal}}{1,000,000/\text{mil}} = 0.0257 \text{ mgd}$$

Next, calculate the MCRT.

Equation: **MCRT, days =**

$$\frac{(\text{MLSS, mg/L})(\text{Aeration tank mil gal} + \text{Clarifier tank mil gal})(8.34 \text{ lb/gal})}{(\text{WAS, mg/L})(\text{Waste rate, mgd})(8.34 \text{ lb/gal}) + (\text{TSS, mg/L})(\text{Flow, mgd})(8.34 \text{ lb/gal})}$$

MCRT, days =

$$\frac{(2,840 \text{ mg/L MLSS})(0.675 \text{ mil gal} + 0.322 \text{ mil gal})(8.34 \text{ lb/gal})}{(6,825 \text{ mg/L})(0.0257 \text{ mgd})(8.34 \text{ lb/gal}) + (18.0 \text{ mg/L TSS})(2.67 \text{ mgd})(8.34 \text{ lb/gal})}$$

$$\text{MCRT, days} = \frac{(2,840 \text{ mg/L MLSS})(0.997 \text{ mil gal})(8.34 \text{ lb/gal})}{1,462.857 \text{ lb/day} + 400.8204 \text{ lb/day}}$$

$$\text{MCRT, days} = \frac{23,614.5432 \text{ lb}}{1,863.6774 \text{ lb/day}} = 12.67 \text{ days, round to } \mathbf{12.7 \text{ days}}$$

DIGESTER VOLATILE SOLIDS LOADING RATIO CALCULATIONS

This calculation compares the volatile solids added to the volatile solids in the digester. It is used for evaluating process control.

172. **What is the ratio of volatile solids loading on a digester, if the digester has 63,750 kg of volatile solids (VS), 1,420 lb/day are pumped into it, percentage total solids (TS) are 5.04%, and percentage volatile solids (VS) are 70.1%?**

First, convert the number of kg to lb.

Know: 1 kg = 2.205 lb

Number of lb = (63,750 kg)(2.205 lb/kg) = 140,568.75 lb

Use the following expanded equation with percentages to solve for digester volatile solid ratio.

Equation: $\textbf{Digester VS ratio} = \dfrac{\textbf{VS added lb/day}}{\textbf{(lb VS in digester)}\,\textbf{(TS\%/100\%)}\,\textbf{(VS\%/100\%)}}$

Substitute values and solve.

$$\text{Digester VS ratio} = \frac{1{,}420\ \text{lb/day}}{(140{,}568.75\ \text{lb})\,(5.04\%/100\%\ \text{TS})\,(70.1\%/100\%\ \text{VS})}$$

$$\text{Digester VS ratio} = \frac{1{,}420\ \text{lb/day}}{(140{,}568.75\ \text{lb})\,(0.0504\ \text{TS})\,(0.701\ \text{VS})} = \textbf{0.286 VS ratio}$$

233

173. **Given the following data, calculate the volatile solids (VS) loading ratio on a digester:**

Sludge weight in digester = 118,375 lb
VS loading = 1,345 lb/day
Total solids (TS) percentage = 5.27%
VS percentage = 69.8%

Use expanded equation with percentages:

$$\textbf{Digester VS ratio} = \frac{\text{VS added lb/day}}{(\text{lb VS in digester})(\text{TS \%}/100\%)(\text{VS \%}/100\%)}$$

Substitute values and solve.

$$\text{Digester VS ratio} = \frac{1,345 \text{ lb/day, VS}}{(118,375 \text{ lb VS})(5.27\%/100\% \text{ TS})(69.8\%/100\% \text{ VS})}$$

$$\text{Digester VS ratio} = \frac{1,345 \text{ lb/day, VS}}{(118,375 \text{ lb VS})(0.0527 \text{ TS})(0.698 \text{ VS})}$$

Digester VS ratio = **0.309 VS ratio**

DIGESTER GAS PRODUCTION PROBLEMS

Operators calculate the amount of gases produced per pound of volatile solids destroyed to determine the effectiveness of the digestion process. Also, it is important to know the gas production because in some cases it is used as a fuel for other plant processes.

174. **If a digester produces 5,475 ft³/day of gas and the amount of volatile solids destroyed are 495 lb/day, what is the amount of gas produced in ft³ per lb of volatile solids (VS) destroyed?**

Equation: **Gas produced, ft³/lb VS destroyed** $= \dfrac{\text{Gas production, ft}^3/\text{day}}{\text{VS destroyed, lb/day}}$

$$\text{Gas produced, ft}^3/\text{lb VS destroyed} = \frac{5,475 \text{ ft}^3/\text{day}}{495 \text{ lb/day}} = \textbf{11.1 ft}^3\textbf{/lb of VS destroyed}$$

175. **What must have been the gas production by a digester in ft³/day given the following data?**

Volatile solids (VS) destroyed = 435 lb/day
Gas produced in ft³/lb VS destroyed = 10.9 ft³/lb

Equation: **Gas produced, ft³/lb VS destroyed** $= \dfrac{\text{Gas production, ft}^3/\text{day}}{\text{VS destroyed, lb/day}}$

Rearrange to solve for gas production in ft³/day.

Gas production, ft³/day = (Gas produced, ft³/lb VS destroyed)(VS destroyed, lb/day)

Substitute values and solve.

Gas production, ft³/day = (10.9 ft³/lb)(435 lb/day) = 4,741.5 ft³/day, round to **4,740 ft³/day**

176. **Given the following data, calculate the gas produced by a digester in cubic meters (m³) per pound of volatile solids (VS) destroyed:**

Digester gas production = 6,145 ft³/day
Volatile solids destroyed = 515 lb/day
1 cubic meter = 35.3 cubic ft

Equation: **Gas produced, m³/lb VS destroyed** $= \dfrac{(\text{Gas production, ft}^3/\text{day})}{(\text{VS destroyed, lb/day})\,(35.3\ \text{m}^3/\text{ft}^3)}$

Gas produced, ft³/lb VS destroyed $= \dfrac{(6,145\ \text{ft}^3/\text{day})}{(515\ \text{lb/day})\,(35.3\ \text{ft}^3/\text{m}^3)} =$ **0.338 m³/lb of VS destroyed**

VOLATILE ACIDS-TO-ALKALINITY RATIO PROBLEMS

The first phase of anaerobic digestion is acid fermentation, which is dependent on new volatile solids entering the digester. The second stage is methane fermentation. These two processes need to be in delicate balance with each other for the anaerobic digestion process to proceed properly. Different treatment plants have different ratios, but typically the ratio is less than 0.1.

177. **Calculate the ratio of volatile acids to alkalinity, if the alkalinity in an anaerobic digester is 1,845 mg/L and the volatile acid concentration of the sludge is 210 mg/L.**

Equation: **Ratio = Volatile acids/Alkalinity**

Substitute values and solve.

$$\text{Ratio} = \frac{210\ \text{mg/L}}{1,845\ \text{mg/L}} = \textbf{0.11 Volatile acids-to-alkalinity ratio}$$

178. **What must have been the volatile acid concentration in an anaerobic digester, if the alkalinity was 2,335 mg/L and the ratio of volatile acids to alkalinity was 0.0958?**

Equation: **Volatile acids = (Alkalinity)(Ratio)**

Substitute values and solve.

Volatile acids = (2,335 mg/L)(0.0958) = 223.69 mg/L, round to **224 mg/L**

179. **What is the amount of volatile solids in lb/day that must have been added to a digester, if the ratio of volatile solids (VS) added to volatile solids already in a digester is 0.085 and the amount of VS already in the digester is 25,400 lb?**

Equation: **Digester VS ratio** $= \dfrac{\text{VS added lb/day}}{\text{lb VS in digester}}$

Rearrange the problem to solve for volatile solids added.

VS added lb/day = (Digester VS ratio)(lb VS in digester)

Substitute values and solve.

VS added lb/day = (0.085 VS ratio)(25,400 lb VS in digester)

VS added lb/day = 2,159 lb/day, round to **2,160 lb/day VS added**

LIME NEUTRALIZATION PROBLEMS

When the sludge in an anaerobic digester becomes acidic it is called a sour digester. A sour digester occurs when the volatile acid-to-alkalinity ratio increases above 0.8. It is not always possible to wait for a digester to naturally correct itself because of the digester's capacity or time constraints. Under these circumstances, it is necessary to neutralize the acid conditions in the digester with lime. The following problems show how operators calculate the appropriate dosage of lime. The lime dosage is based on the amount of volatile acids in the sludge and is in a 1-to-1 ratio, that is, 1 mg/L of lime will neutralize 1 mg/L of volatile acid.

180. **How many pounds of lime are required to neutralize a sour digester that contains 165,000 gallons of sludge and a volatile acid content of 2,010 mg/L?**

Know: 1 mg/L of lime will neutralize 1 mg/L of volatile acids.

First, convert the digester's volume from gallons to mil gal.

$$\text{Number of mil gal} = \frac{165,000 \text{ gal}}{1,000,000/\text{mil}} = 0.165 \text{ mil gal}$$

Equation: **Number of lb = (Volatile acids, mg/L)(mil gal)(8.34 lb/gal)**

Substitute values and solve.

Lime, lb = (2,010 mg/L)(0.165 mil gal)(8.34 lb/gal) = 2,765.96 lb, round to **2,770 lb of Lime**

181. **Given the following data, calculate the amount of lime in lb that are needed to neutralize a sour digester.**

Digester volume = 240,750 gallons
Volatile acids = 2,245 mg/L

Know: 1 mg/L of lime will neutralize 1 mg/L of volatile acids.

First, convert the digester's volume from gallons to mil gal.

$$\text{Number of mil gal} = \frac{240,750 \text{ gal}}{1,000,000/\text{mil}} = 0.24075 \text{ mil gal}$$

Equation: **Number of lb = (Volatile acids, mg/L)(mil gal)(8.34 lb/gal)**

Substitute values and solve.

Lime, lb = (2,245 mg/L)(0.24075 mil gal)(8.34 lb/gal) = 4,507.6 lb, round to **4,510 lb of Lime**

182. **What must have been the concentration of volatile acids in mg/L for a sour digester with a volume of 175,000 gallons, if the number of pounds of lime to neutralize the volatile acids was 1,975 lb?**

Know: 1 mg/L of lime will neutralize 1 mg/L of volatile acids.

First, convert the digester's volume from gallons to mil gal.

$$\text{Number of mil gal} = \frac{175,000 \text{ gal}}{1,000,000/\text{mil}} = 0.175 \text{ mil gal}$$

Equation: **Number of lb lime = (Volatile acids, mg/L)(mil gal)(8.34 lb/gal)**

Rearrange the equation to solve for the concentration of volatile acids.

$$\text{Volatile acids, mg/L} = \frac{\text{Number of lb, lime}}{(\text{mil gal})(8.34 \text{ lb/gal})}$$

Substitute values and solve.

$$\text{Volatile acids, mg/L} = \frac{1,975 \text{ lb, lime}}{(0.175 \text{ mil gal})(8.34 \text{ lb/gal})}$$

Volatile acids, mg/L = 1,353.2 mg/L, round to **1,350 mg/L Volatile acids**

POPULATION LOADING CALCULATIONS

These calculations are used for wastewater treatment ponds. They are based on the number of people per acre of pond, and it is a helpful tool in evaluating process control of ponds.

183. **Given the following data, calculate the population loading in people per acre on two ponds:**

Pond one = 4.63 acres
Pond two = 5.82 acres
Services = 6,895
Average persons per service = 3.18

First, add the area in acres for each pond to get the total acres.

Total area of ponds = 4.63 acres + 5.82 acres = 10.45

Next, determine the number of people served.

Number of people = (6,895 services)(3.18 people/service) = 21,926 people served

Next, using the following equation, determine the population loading.

$$\text{\textbf{Population loading, people/acre}} = \frac{\text{Number of people served}}{\text{Area of pond(s), acres}}$$

Substitute values and solve.

$$\text{Population loading, people/acre} = \frac{21{,}926 \text{ people served}}{10.45 \text{ acres}}$$

Population loading, people/acre = 2,098 people/acre, round to **2,100 people/acre**

184. **What is the population loading in people/acre, if there are 35,300 services with 3.10 people per service, given the following data?**

Wastewater pond 1 = 10.86 acres
Wastewater pond 2 = 8.23 acres
Wastewater pond 3 = 9.44 acres
Wastewater pond 4 = 8.17 acres

First, add the area in acres for each pond to get the total acres.

Total area of ponds = 10.86 acres + 8.23 acres + 9.44 acres + 8.17 acres = 36.7 acres

Next, determine the number of people served.

Number of people = (35,300 services)(3.10 people/service) = 109,430 people served

Next, using the following equation determine the population loading.

$$\textbf{Population loading, people/acre} = \frac{\text{Number of people served}}{\text{Area of pond(s), acres}}$$

Substitute values and solve.

$$\text{Population loading, people/acre} = \frac{109{,}430 \text{ people served}}{36.7 \text{ acres}}$$

Population loading, people/acre = 2,981.74 people/acre, round to **2,980 people/acre**

POPULATION EQUIVALENT CALCULATIONS

Wastewater discharge from industries or commercial sources usually has a higher organic content than domestic wastewaters. Operators use population equivalent calculations to compare domestic wastewater to wastewater from these former sources. This is important in determining the loading that will be placed on a wastewater system when a new industry wants to connect to a system. What is needed is the flow from this industry in mgd and the BOD_5 concentration in mg/L. Domestic wastewater systems usually contain a range of 0.17 to 0.20 pounds of BOD_5 per day, which the wastewater plant should have already determined. Also, population equivalent calculations are required for designing proper size wastewater treatment plants, pump stations, and pipe sizes, because the volumetric flow that is expected to be treated and pumped needs to be estimated.

185. **Given the following data, calculate the population equivalent:**

Average wastewater flow for the day = 3.42 ft³/s
BOD_5/person = 0.26 lb/day/person
BOD_5 concentration in the wastewater = 2,317 mg/L

First, convert ft³/s to mgd.

Number of mgd = (3.42 ft³/s)(0.6463 mgd/ ft³/s) = 2.21 mgd

Use the following equation to solve this problem.

$$\textbf{Number of people} = \frac{(BOD_5, \text{mg/L})(\text{mgd})(8.34 \text{ lb/gal})}{\text{lb/day of BOD/person}}$$

$$\text{Number of people} = \frac{(2,317 \text{ mg/L } BOD_5)(2.21 \text{ mgd})(8.34 \text{ lb/gal})}{0.26 \text{ lb/day/person}}$$

Number of people = 164,252 people, round to **160,000 people**

186. **A wastewater treatment plant has an influent flow of 2,808,000 gallons. If the BOD$_5$ is 2,470 mg/L and the average BOD$_5$ per person is 0.29 lb/day, what is the population equivalent that this plant is currently treating?**

First, convert the flow in gallons to mgd.

$$\text{Number of mgd} = \frac{2,808,000 \text{ gal}}{1,000,000/\text{mil}} = 2.808 \text{ mgd}$$

Next, use the following equation to solve this problem.

$$\textbf{Number of people} = \frac{(\text{BOD}_5, \text{mg/L})(\text{mgd})(8.34 \text{ lb/gal})}{\text{lb/day of BOD}_5/\text{person}}$$

$$\text{Number of people} = \frac{(2,470 \text{ mg/L BOD}_5)(2.808 \text{ mgd})(8.34 \text{ lb/gal})}{0.29 \text{ lb/day}}$$

Number of people = 199,463 people, round to **200,000 people**

SOLIDS UNDER AERATION

Solids under aeration calculations are used by operators for evaluating process control.

187. **Calculate the number of pounds of suspended solids (SS) contained in an aeration tank, if the tank contains 214,000 gallons, the concentration of SS is 2,045 mg/L, and the specific gravity is 1.05.**

First, convert gallons to mil gal.

$$\text{Number of mil gal} = \frac{214,000 \text{ gal}}{1,000,000/\text{mil}} = 0.214 \text{ mil gal}$$

Next, determine the number of lb/gal for the SS.

Number of lb/gal, SS = (8.34 lb/gal)(1.05 sp gr) = 8.757 lb/gal

Next, determine the pounds of SS under aeration using a modified version of the "pounds" equation because the SS weighs more than water.

Equation: **Number of lb SS = (SS, mg/L)(Number of mil gal)(SS lb/gal)**

Substitute values and solve.

Number of lb SS = (2,045 mg/L SS)(0.214 mil gal)(8.757 lb/gal)

Number of lb SS = 3,832.33 lb, round to **3,830 lb SS**

188. **How many pounds of mixed liquor suspended solids (MLSS) are being aerated, if the aeration tank is 51.2 ft in diameter, with a sludge height of 12.9 ft, the concentration of MLSS is 1,988 mg/L, and the specific gravity of the MLSS is 1.03?**

First, determine how many gallons are in the aeration tank.

Number of gallons = (0.785)(Diameter)^2(Height, ft)(7.48 gal/ft^3)

Number of gallons = (0.785)(51.2 ft)(51.2 ft)(12.9 ft)(7.48 gal/ft^3) = 198,564 gal

Next, convert gallons to mil gal.

Number of mil gal $= \dfrac{198,564 \text{ gal}}{1,000,000/\text{mil}} = 0.198564$ mil gal

Next, determine the lb/gal for the MLSS.

Number of lb/gal, MLSS = (8.34 lb/gal)(1.03 sp gr) = 8.59 lb/gal

Next, determine the pounds of MLSS under aeration using a modified version of the "pounds" equation because the MLSS weighs more than water (8.34 lb/gal).

Equation: **Number of lb MLSS = (MLSS, mg/L)(Number of mil gal)(MLSS lb/gal)**

Substitute values and solve.

Number of lb MLSS = (1,988 mg/L MLSS)(0.198564 mil gal)(8.59 lb/gal)

Number of lb MLSS = 3,390.86 lb, round to **3,390 lb MLSS**

SUSPENDED SOLIDS REMOVAL

The suspended solids removal calculations are used by operators as a sign for the efficiency of the treatment process in question. Typically, the suspended solids removed from wastewater systems ranges from 100 to 350 mg/L.

189. **Given the following data, calculate the quantity of suspended solids (SS) in lb/day that was removed from a primary clarifier:**

Average influent flow for the day = 4.89 ft³/s
Suspended solids = 107 mg/L
Specific gravity of the SS = 1.02

First, convert ft³/s to mgd.

Number of mgd = (4.89 ft³/s)(0.6463 mgd/ ft³/s) = 3.16 mgd

Next, determine the lb/gal for the SS.

Number of lb/gal, SS = (8.34 lb/gal)(1.02 sp gr) = 8.51 lb/gal

Next, determine the pounds of SS under aeration using a modified version of the "pounds" equation because the SS weighs more than water.

Equation: **Number of lb/day SS = (SS, mg/L)(Number of mgd)(SS lb/gal)**

Substitute values and solve.

Number of lb/day SS removed = (107 mg/L SS)(3.16 mgd)(8.51 lb/gal)

Number of lb/day SS removed = 2,877.40 lb/day, round to **2,880 lb/day SS removed**

190. **What must have been the average influent concentration of suspended solids (SS) in mg/L, if a wastewater treatment plant's clarifier had a flow of 1,385,000 gpd and the clarifier removed 2,270 lb/day of SS?**

First, convert gpd to mgd.

$$\text{Number of mgd} = \frac{1,385,000 \text{ gpd}}{1,000,000/\text{mil}} = 1.385 \text{ mgd}$$

Equation: **Number of lb/day SS = (SS, mg/L)(Number of mgd)(8.34 lb/gal)**

Rearrange the equation, then substitute values and solve.

$$\textbf{SS removed, mg/L} = \frac{\text{Number of lb/day SS}}{(\text{Number of mgd})(8.34 \text{ lb/gal})}$$

SS removed, mg/L $= \dfrac{2,270 \text{ lb/day}}{(1.385 \text{ mgd})(8.34 \text{ lb/gal})} = 196.52$ mg/L, round to **197 mg/L SS removed**

BIOCHEMICAL OXYGEN DEMAND REMOVAL CALCULATIONS

The biochemical oxygen demand (BOD_5) removal calculations are used to inform operators about the efficiency of the treatment process for a pond or trickling filter. The BOD_5 is an empirical test that informs the operator on the relative oxygen requirements of a wastewater and is an indicator of how much food is in the wastewater. The BOD_5 is a 5-day test.

191. **Given the following data, determine the BOD_5 removal in lb/day from a trickling filter.**

Plant influent flow = 1.71 mgd
Influent BOD_5 concentration = 294 mg/L
Effluent BOD_5 concentration = 108 mg/L

First, determine the amount of BOD_5 removed in mg/L by subtracting the influent BOD_5 from the effluent BOD_5.

BOD_5 removed, mg/L = Influent BOD_5, mg/L − Effluent BOD_5, mg/L

BOD_5 removed, mg/L = 294 mg/L − 108 mg/L = 186 mg/L BOD_5 removed

Next, solve the amount of BOD_5 removed in lb/day by using the "pounds" equation.

Equation: **Number of lb/day BOD_5 = (BOD_5, mg/L)(Number of mgd)(8.34 lb/gal)**

Substitute values and solve.

Number of lb/day BOD_5 removed = 186 mg/L BOD_5)(1.71 mgd)(8.34 lb/gal

Number of lb/day BOD_5 removed = 2,652.62 lb/day, round to **2,650 lb/day BOD_5 removed**

192. **What must have been the daily flow to a trickling filter in mgd, if the influent BOD_5 was 347 mg/L, the effluent BOD_5 was 126 mg/L, and the BOD_5 removed was 4,165 lb/day?**

First, the amount of BOD_5 removed must still be determined by subtracting the influent BOD_5 from the effluent BOD_5.

BOD_5 removed, mg/L = 347 mg/L − 126 mg/L = 221 mg/L BOD_5 removed

Next, solve the amount of BOD_5 removed in lb/day by using the "pounds" equation

Equation: **Number of lb/day BOD_5 = (BOD_5, mg/L)(Number of mgd)(8.34 lb/gal)**

Rearrange the equation, then substitute values and solve.

$$\textbf{Number of mgd} = \frac{\text{Number of lb/day BOD}_5}{(\text{BOD}_5 \text{ removed, mg/L})(8.34 \text{ lb/gal})}$$

Substitute values and solve.

$$\text{Number of mgd} = \frac{4,165 \text{ lb/day}}{(221 \text{ mg/L BOD}_5 \text{ removed})(8.34 \text{ lb/gal})} = \textbf{2.26 mgd}$$

193. What is the BOD_5 removal in lb/day from a 2.85 mgd wastewater plant, if the population equivalent of BOD_5 is 116 mg/L and the secondary BOD_5 from the effluent is 23 mg/L?

First, determine the BOD_5 removal in mg/L.

BOD_5 removal, mg/L = 116 mg/L − 23 mg/L = 93 mg/L

Next, calculate the BOD_5 removal in lb/day.

Equation: **BOD_5 removal, lb/day = (Removed BOD_5, mg/L)(mgd)(8.34 lb/gal)**

BOD_5 removal, lb/day = (93 mg/L)(2.85 mgd)(8.34 lb/gal)

BOD_5 removal, lb/day = 2,210.5 lb/day, round to **2,200 lb/day BOD_5 removal**

FOOD-TO-MICROORGANISM RATIO CALCULATIONS

A properly operated activated sludge process has a balance between the food entering the system and the microorganisms in the aeration tank. The best ratio varies because it depends on the activated sludge process and the characteristics of the wastewater being treated. The calculation is a measure of the pounds of food coming in divided by the pounds of microorganisms present. The ratio is a process control number because it helps the operator determine the proper number of microorganisms for the system in question.

194. What is the food-to-microorganism (F/M) ratio for an aeration tank that is 49.9 ft in diameter, with a liquid level of 13.5 ft, if the primary effluent flow averages 3,540,000 gpd, the mixed liquor volatile suspended solids (MLVSS) is 2,912 mg/L, and the BOD_5 is 302 mg/L?

First, calculate the number of gallons in the aeration tank.

Number of gallons = (0.785)(Diameter)^2(Height, ft)(7.48 gal/ft^3)

Number of gallons = (0.785)(49.9 ft)(49.9 ft)(13.5 ft)(7.48 gal/ft^3) = 197,381 gal

Next, convert gallons to mil gal.

$$\text{Number of mil gal} = \frac{197,381 \text{ gpd}}{1,000,000/\text{mil}} = 0.197381 \text{ mil gal}$$

Next, convert the effluent flow in gpd to mgd.

Number of mgd = 3,540,000 gal/1,000,000/mil = 3.54 mgd

Next, write the equation.

$$\text{F/M} = \frac{(\text{BOD}_5, \text{mg/L})(\text{Flow, mgd})}{(\text{mg/L MLVSS})(\text{Volume of tank, mil gal})}$$

Substitute values and solve.

$$\text{F/M} = \frac{(302 \text{ mg/L BOD}_5)(3.54 \text{ mgd})}{(2,912 \text{ mg/L MLVSS})(0.197381 \text{ mil gal})} = \textbf{1.86 F/M ratio}$$

195. Given the following data on an aeration tank, calculate the current food-to-microorganism (F/M) ratio.

Primary effluent flow = 3.64 mgd
Volume of aeration tank = 239,000 gallons
Mixed liquor volatile suspended solids (MLVSS) = 2,888 mg/L
BOD$_5$ = 231 mg/L

First, convert the volume of wastewater in the aeration tank to mil gal.

$$\text{Number of mil gal} = \frac{239,000 \text{ gal}}{1,000,000/\text{mil}} = 0.239 \text{ mil gal}$$

Next, write the equation: $\text{F/M} = \dfrac{(\text{BOD}_5, \text{mg/L})(\text{Flow, mgd})(8.34 \text{ lb/gal})}{(\text{mg/L MLVSS})(\text{Volume in tank, mil gal})(8.34 \text{ lb/gal})}$

The 8.34 lb/gal cancels leaving the following equation:

$$\text{F/M} = \frac{(\text{BOD}_5, \text{mg/L})(\text{Flow, mgd})}{(\text{mg/L MLVSS})(\text{Volume of tan k, mil gal})}$$

Substitute values and solve.

$$\text{F/M} = \frac{(231 \text{ mg/L BOD}_5)(3.64 \text{ mgd})}{(2,888 \text{ mg/L MLVSS})(0.239 \text{ mil gal})} = \textbf{1.22 F/M ratio}$$

SEED SLUDGE PROBLEMS

This calculation is required for determining how much seed sludge in gallons to use for starting a new digester.

196. **A digester with a diameter of 40.2 ft and a sludge level of 16.8 ft has a seed sludge requirement of 13.5% of the digester capacity. How many gallons of seed sludge will be needed?**

First, determine the number of gallons in the digester.

Volume, gal = (0.785)(Diameter)^2(Height, ft)(7.48 gal/ft^3)

Volume, gal = (0.785)(40.2 ft)(40.2 ft)(16.8 ft)(7.48 gal/ft^3) = 159,416 gal

Next, use the following equation.

$$\textbf{Seed sludge, gal} = \frac{\text{(Capacity of digester)(Percent seed sludge required)}}{100\%}$$

$$\text{Seed sludge, gal} = \frac{(159,416 \text{ gallons})(13.5\%)}{100\%} = \textbf{21,500 gal of seed sludge}$$

197. **Given the following data, determine the seed sludge required in gallons:**

Digester has a radius of 24.9 ft
Liquid level in digester is 18.6 ft
Requires 16.0% seed sludge

First, determine the number of gallons in the digester.

Volume, gal = πr^2(Depth, ft)(7.48 gal/ft^3)

Volume, gal = (3.14)(24.9 ft)(24.9 ft)(18.6 ft)(7.48 gal/ft^3) = 270,859 gal

Next, use the following equation.

$$\textbf{Seed sludge, gal} = \frac{(\text{Capacity of digester})(\text{Percent seed sludge required})}{100\%}$$

$$\text{Seed sludge, gal} = \frac{(270,859 \text{ gallons})(16.0\%)}{100\%}$$

Seed sludge, gal = 43,337 gal, round to **43,300 gal of seed sludge**

GRAVITY THICKENER SOLIDS LOADING PROBLEMS

Gravity thickeners use large tanks that separate suspended solids and mineral matter from the liquid by gravity. The gravity thickener concentrates the sludge to reduce the load on processes that follow—conditioning, dewatering and digestion—and produces a clear liquid, which is decanted. Flocculants are used to speed up the settling process. Operators can calculate the solids loading in lb/d/ft^2 or the hydraulic loading in gal/day/ft^2. The hydraulic loading calculation is used by operators to determine if the process is being overloaded or underloaded. See Figures 4, 5, and 6 in Appendix E for three types of wastewater plants using the thickening process.

198. **Given the following data, determine the solids loading on a gravity thickener in lb/d/ft^2:**

Gravity thickener = 48.1 ft in diameter
Influent flow = 36.6 gpm
Percent solids = 3.25%

First, determine the area of the gravity thickener.

Know: Area = (0.785)(Diameter)^2

Area = (0.785)(48.1 ft)(48.1 ft) = 1,816.18 ft^2

Equation: **Solids loading, lb/d/ft^2 =**

$$\frac{(\text{Flow, gpm})(1,440 \text{ min/day})(\text{Percent solids})(8.34 \text{ lb/gal})}{(\text{Gravity thickener area})(100\%)}$$

Substitute values and solve.

$$\text{Solids loading, lb/d/ft}^2 = \frac{(36.6 \text{ gpm})(1,440 \text{ min/day})(3.25\%)(8.34 \text{ lb/gal})}{(1,816.18 \text{ ft}^2)(100\%)}$$

Solids loading, lb/d/ft^2 = 7.8656 lb/d/ft^2, round to **7.87 lb/d/ft^2**

199. What are the solids loading in lb/d/ft^2 on a gravity thickener, if the percent solids is 4.85%, the influent flow is 53 gpm, and the gravity thickener has a radius of 29.9 ft?

First, convert the gpm to gpd.

Know: Area of gravity thickener = πr^2 where π = 3.14

Equation: **Solids loading, lb/d/ft^2** $= \dfrac{(\text{Flow, gpd})(1,440 \text{ min/day})(8.34 \text{ lb/gal})(\text{Percent solids})}{(\text{Gravity thickener area})(100\%)}$

Substitute values and solve.

$$\text{Solids loading, lb/d/ft}^2 = \frac{(53 \text{ gpm})(1,440 \text{ min/day})(8.34 \text{ lb/gal})(4.85\%)}{(3.14)(29.9 \text{ ft})(29.9 \text{ ft})(100\%)}$$

Solids loading, lb/d/ft^2 = 10.997 lb/d/ft^2, round to **11.0 lb/d/ft^2**

DISSOLVED AIR FLOTATION: THICKENER SOLIDS LOADING PROBLEMS

The dissolved air flotation (DAF) technique is used to thicken sludge. These types of calculations are used for evaluating process control.

200. **What are the solids loading for a DAF unit in lb/hr/ft² that is 54.1 ft by 18.3 ft, that has a sludge flow of 0.179 mgd and a waste-activated sludge (WAS) concentration of 6,181 mg/L, and the sludge has a specific gravity of 1.05?**

First, determine the area of the DAF unit in ft².

DAF area, ft² = (54.1 ft)(18.3 ft) = 990.03 ft²

Next, calculate the weight of the sludge in lb/gal.

Sludge, lb/gal = (8.34 lb/gal)(1.05 sp gr) = 8.757 lb/gal

Next, calculate the solids loading.

Equation: **Solids loading, lb/hr/ft²** $= \dfrac{(\text{WAS, mg/L}) \, (\text{mgd}) \, (\text{lb/gal, Sludge})}{(\text{DAF area, ft}^2) \, (24 \text{ hr/day})}$

Solids loading, lb/hr/ft² $= \dfrac{(6,181 \text{ mg/L, WAS}) \, (0.179 \text{ mgd}) \, (8.757 \text{ lb/gal, Sludge})}{(990.03 \text{ ft}^2 \text{ DAF}) \, (24 \text{ hr/day})}$

Solids loading, lb/hr/ft² = 0.4078 lb/hr/ft², round to **0.408 lb/hr/ft²**

201. **Given the following data, calculate the solids loading in lb/d/ft² on a dissolved air flotation (DAF) thickener unit:**

Area of DAF = 1,375 ft²
Waste-activated sludge (WAS) = 8,110 mg/L
Sludge flow = 125 gpm
Sludge specific gravity = 1.04 lb/gal

First, convert gpm to mgd.

$$\text{Number of gpd} = \frac{(125\ \text{gpm})\,(1{,}440\ \text{min}/\text{day})}{1{,}000{,}000/\text{mil}} = 0.18\ \text{mgd}$$

Next, determine the lb/gal for the sludge using the specific gravity.

$$\text{Sludge, lb/gal} = (8.34\ \text{lb/gal})(1.04\ \text{sp gr}) = 8.6736\ \text{lb/gal}$$

Equation: **Solids loading, lb/d/ft²** $= \dfrac{(\text{WAS, mg/L})\,(\text{mgd})\,(\text{lb/gal, Sludge})}{\text{DAF area, ft}^2}$

$$\text{Solids loading, lb/d/ft}^2 = \frac{(8{,}110\ \text{mg/L, WAS})\,(0.18\ \text{mgd})\,(8.6736\ \text{lb/gal, Sludge})}{1{,}375\ \text{ft}^2\ \text{DAF}}$$

Solids loading, lb/d/ft² $= 9.209$ lb/d/ft², round to **9.21 lb/d/ft²**

DISSOLVED AIR FLOTATION: AIR-TO-SOLIDS RATIO CALCULATIONS

Air-to-solids ratio calculations are used to determine the efficiency of the process, as the air flotation thickener and the solids in the system must be in balance. Typically the ratio ranges from 0.01 to 0.1.

202. **What is the air-to-solids ratio for a dissolved air flotation (DAF) unit that has an air flow rate of 6.8 ft³/min, a solids concentration of 0.70%, and a flow of 127,000 gpd?**

Know: Air $= 0.0807$ lb/ft³ at standard temperature, pressure, and average composition

Equation: **Air-to-solids ratio** $= \dfrac{(\text{Air flow, ft}^3/\text{min}\,)\,(\text{Air, lb/ft}^3)}{(\text{gpm})\,(\text{Percent solids}/100\%)\,(8.34\ \text{lb/gal})}$

Since equation requires gpm, first convert gpd to gpm.

$$\text{Number of gpm} = \frac{127{,}000\ \text{gpd}}{1{,}440\ \text{min}/\text{day}} = 88.19\ \text{gpm}$$

Now, using the air-to-solids equation, substitute values and solve.

$$\text{Air-to-solids ratio} = \frac{(6.8\ \text{ft}^3/\text{min})\,(0.0807\ \text{lb/ft}^3)}{(88.19\ \text{gpm})\,(0.70\%/100\%)\,(8.34\ \text{lb/gal})} = \textbf{0.11 Air-to-solids ratio}$$

203. Given the following data, determine the air-to-solids ratio for a DAF unit.

DAF influent flow = 115,000 gpd
Air flow = 7.35 ft³/min
Solids concentration = 0.775%
Solids specific gravity = 1.06

Know: Air = 0.0807 lb/ft³ at standard temperature, pressure, and average composition

Equation: **Air-to-solids ratio** $= \dfrac{(\text{Air flow, ft}^3/\min)\,(\text{Air, lb/ft}^3)}{(\text{gpm})\,(\text{Percent solids}/100\%)\,(\text{Solids, lb/gal})}$

Since equation requires gpm, first convert gpd to gpm.

Number of gpm $= \dfrac{115,000 \text{ gpd}}{1,440 \text{ min/day}} = 79.86 \text{ gpm}$

Then, the number of lb/gal is required since the solids weigh more than water.

Solids, lb/gal = (8.34 lb/gal)(1.06 sp gr) = 8.84 lb/gal

Now, using the air-to-solids equation, substitute values and solve.

Air-to-solids ratio $= \dfrac{(7.35 \text{ ft}^3/\min)\,(0.0807 \text{ lb/ft}^3)}{(79.86 \text{ gpm})\,(0.775\%/100\%)\,(8.84 \text{ lb/gal})} = $ **0.108 Air-to-solids ratio**

DISSOLVED AIR FLOTATION: AIR RATE FLOW CALCULATIONS

Operators use air rate flow calculations for evaluating process control.

204. If a DAF unit receives air at an average rate of 0.285 m³/min, how many lb/day of air does it receive? Note: 1 cubic meter = 35.3 ft³.

First, convert m³/min to ft³/min.

Air flow, ft³/min = (0.285 m³/min)(35.3 ft³/m³) = 10.06 ft³/min

Know: Air $= 0.0807$ lb/ft^3 at standard temperature (4°C at "0°" meridian and at sea level), pressure, and average composition

Equation: **Air, lb/day $=$ (Air flow, ft^3/min)(1,440 min/day)(0.0807 lb/ft^3, Air)**

Substitute values and solve.

Air, lb/day $=$ (10.06 ft^3/min)(1,440 min/day)(0.0807 lb/ft^3)

Air, lb/day $= 1,169.05$ lb/day, round to **1,170 lb/day of Air**

205. If a DAF unit receives air at an average rate of 0.208 m^3/min, how many lb/hr of air does it receive?

First, convert m^3/min to ft^3/min.

Air flow, ft^3/min $=$ (0.208 m^3/min)(35.3 ft^3/m^3) $= 7.34$ ft^3/min

Know: Air $= 0.0807$ lb/ft^3 at standard temperature, pressure, and average composition

Equation: **Air, lb/day $=$ (Air flow, ft^3/min)(60 min/hr)(0.0807 lb/ft^3, Air)**

Substitute values and solve.

Air, lb/day $=$ (7.34 ft^3/min)(60 min/hr)(0.0807 lb/ft^3)

Air, lb/day $= 35.54$ lb/hr, round to **35.5 lb/hr of Air**

CENTRIFUGE THICKENING PROBLEMS

Centrifuges are used to dewater sludge usually after applying gravity thickening. They apply forces that are a thousand times greater than gravity. Polymers may be applied to the influent of the centrifuge to facilitate solids thickening.

206. Given the following data, determine the removal efficiency of sludge solids on a disc centrifuge.

Influent sludge solids = 7,125 mg/L
Effluent sludge solids = 612 mg/L

First, determine the amount of sludge solids removed.

Sludge solids removed = 7,125 mg/L − 612 mg/L = 6,513 mg/L

Now, determine the removal efficiency.

$$\text{Percent removal efficiency} = \frac{\text{Solids removed, mg/L}}{\text{Influent solids, mg/L}}$$

$$\text{Percent removal efficiency} = \frac{(6,513 \text{ mg/L})(100\%)}{7,125 \text{ mg/L}} = 91.41\%, \text{ round to } \mathbf{91.4\%}$$

207. Given the following data, what is the feed time for a basket centrifuge thickener?

Basket centrifuge thickener capacity = 25 ft³
Sludge flow rate = 50,400 gpd
Solids concentration = 6,875 mg/L
Percent solids = 6.15%

Know: 1% = 10,000 ppm or mg/L

First, convert the solids concentration in mg/L to percent

$$\text{Percent solids} = \frac{6,875 \text{ mg/L}}{10,000 \text{ mg/L/1\%}} = 0.6875\%$$

Next, convert gpd to gpm.

$$\text{Number of gpm} = \frac{50,400 \text{ gpd}}{1,440 \text{ min/day}} = 35 \text{ gpm}$$

Now, calculate the feed time in minutes.

Equation: $\textbf{Feed time, min} = \dfrac{(\text{Capacity, ft}^3)(\text{Solids, \%}/100\%)(7.48 \text{ gal/ft}^3)(8.34 \text{ lb/gal})}{(\text{Flow, gpm})(\text{Solids concentration, \%}/100\%)(8.34 \text{ lb/gal})}$

Simplify the equation by canceling out the 8.34 lb/gal and the 100%

$$\textbf{Feed time, min} = \frac{(\text{Capacity, ft}^3)(\text{Solids, \%})(7.48 \text{ gal/ft}^3)}{(\text{Flow, gpm})(\text{Solids concentration, \%})}$$

$$\text{Feed time, min} = \frac{(25 \text{ ft}^3)(6.15\%)(7.48 \text{ gal/ft}^3)}{(35 \text{ gpm})(0.6875\%)} = 47.79 \text{ min, round to } \textbf{48 min}$$

SAND DRYING BED PROBLEMS

By knowing how thick the sludge was when applied and later measuring the thickness of the dried sludge, an operator can use these calculations to determine the efficiency of the drying bed process.

208. A drying bed is 285 ft long and 49.5 ft wide. If 4.5 inches of sludge were applied to the drying bed, how many gallons of sludge were applied?

First, convert 4.5 in. to feet.

Number of feet = 4.5 in./12 in./ft = 0.375 ft

Next, determine the volume in ft³ sent to the drying bed.

Volume, ft³ = (285 ft)(49.5 ft)(0.375 ft) = 5,290.3 ft³

Lastly, calculate the volume in gallons sent to the sand drying beds.

Number of gal = (5,290.3 ft³)(7.48 gal/ft³) = 39,571.44 gal, round to **40,000 gal**

209. **A sand drying bed is 220 ft long and 45 ft wide. If a digester that is 59.9 ft in diameter is drawn down by 2.75 ft, how many gallons of sludge will be sent to the sand drying beds?**

Equation: **Volume, gal $= (0.785)(\text{Diameter, ft})^2(\text{Depth, ft})(7.48 \text{ gal/ft}^3)$**

Volume, gal $= (0.785)(59.9 \text{ ft})(59.9 \text{ ft})(2.75 \text{ ft})(7.48 \text{ gal/ft}^3) = 57{,}937$ gal, round to **57,900 gal**

DEWATERING CALCULATIONS

This section contains several types of dewatering problems. The more water removed from sludge the less cost associated with further processing or disposal. The problems are important to the operator because they are helpful in evaluating process control or in informing the operator of process efficiency. See Figure 12 in Appendix E for one type of sludge process using dewatering.

210. **What is the sludge feed rate for a belt filter press to process 9,875 lb/day of sludge, if it operates only 8.50 hr/day?**

Sludge feed rate, lb/hr $= \dfrac{9{,}875 \text{ lb/day}}{8.50 \text{ hr/day}} = 1{,}161.76$ lb/hr, round to **1,160 lb/hr**

211. **If the maximum feed rate to a belt filter press for an adequate cake is 1,350 lb/hr, how long in hours will it take to process 78,500 lb of sludge?**

Number of hours $= \dfrac{78{,}500 \text{ lb}}{1{,}350 \text{ lb/hr}} =$ **58.1 hr**

212. **A vacuum filter has a wet cake flow of 4,515 lb/hr and a filter that is 14.9 by 24.1 ft. Calculate the filter yield in lb/hr/ft², if the percent solids are 21.5%.**

First, determine the area of the filter.

Area of filter, ft² $= (14.9 \text{ ft})(24.1 \text{ ft}) = 359.09 \text{ ft}^2$

Equation: **Filter yield, lb/hr/ft²** $= \dfrac{(\text{Wet cake flow, lb/hr})(\text{Percent solids}/100\%)}{\text{Area, ft}^2}$

Substitute values and solve.

Filter yield, lb/hr/ft² $= \dfrac{(4{,}515 \text{ lb/hr})\,(21.5\%/100\%)}{359.09 \text{ ft}^2} =$ **2.70 lb/hr/ft²**

213. **Calculate the filter loading in lb/d/ft² on a vacuum filter that has a diameter of 8.25 ft and a length of 29.9 ft, if the digested biosolids flow rate is 57 gpm and it has a solids concentration of 4.35%.**

First, convert gpm to gpd.

Number of gpd $=$ (57 gpm)(1,440 min/day) $=$ 82,080 gpd

Next, calculate the area of the vacuum filter in ft².

Know: If the filter is unwrapped from the drum and laid flat the width would simply be π(Diameter), the circumference, where π is equal to 3.14.

Equation: **Area, ft² = (Length, ft)(Width, ft)**, where the width $= \pi$(Diameter)

Area, ft² $=$ (29.9 ft)(3.14)(8.25 ft) $=$ 774.56 ft²

Next, solve for filter loading.

Equation: **Vacuum filter loading, lb/d/ft²** $= \dfrac{(\text{Biosolids, gpd})\,(8.34 \text{ lb/gal})\,(\text{Percent solids})}{\text{Vacuum filter area, ft}^2}$

Vacuum filter loading, lb/d/ft² $= \dfrac{(82{,}080 \text{ gpd})\,(8.34 \text{ lb/gal})\,(4.35\%/100\%)}{774.56 \text{ ft}^2}$

Vacuum filter loading, lb/d/ft² $=$ 38.44 lb/d/ft², round to **38 lb/d/ft²**

214. **Given the following data, calculate the filter loading in lb/d/ft² on a vacuum filter that has a diameter of 9.35 ft and a length of 32.1 ft.**

Digested biosolids flow rate = 52 gpm
Solids concentration = 3.95%

First, convert gpm to gal/day.

Number of gpd = (52 gpm)(1,440 min/day) = 74,880 gpd

Next, calculate the area of the vacuum filter in ft².

Know: As above the width will be π(Diameter), the circumference, where π is equal to 3.14.

Equation: **Area, ft² = (Length, ft)π(Diameter, ft)**

Area, ft² = (32.1 ft)(3.14)(9.35 ft) = 942.42 ft²

Next, solve for filter loading.

Equation: **Vacuum filter loading, lb/d/ft²** $= \dfrac{(\text{Biosolids, gpd})(8.34\ \text{lb/gal})(\text{Percent solids})}{\text{Vacuum filter area, ft}^2}$

Vacuum filter loading, lb/d/ft² $= \dfrac{(74,880\ \text{gpd})(8.34\ \text{lb/gal})(3.95\%/100\%)}{942.42\ \text{ft}^2}$

Vacuum filter loading, lb/d/ft² = 26.17 lb/d/ft², round to **26 lb/d/ft²**

VACUUM FILTER YIELD AND OPERATING TIME CALCULATIONS

The vacuum filter yield calculation is used to measure performance. The operating time is the time required to process the solids and is used for planning purposes.

215. **A vacuum filter with a surface area of 218 ft², processes an average of 2,390 lb/ day. If the solids recovery averages 94.2% and the vacuum filter yield averages 1.93 lb/hr/ft, how many hours per day will it take the vacuum filter to process these solids?**

Equation: **Filter yield, lb/hr/ft²** $= \dfrac{\dfrac{(\text{Solids, lb/day})(\text{Percent recovery})}{(\text{Filter operation, lb/day})(100\%)}}{\text{Filter area ft}^2}$

Substitute values and solve.

$$1.93 \text{ lb/hr/ft}^2 = \dfrac{\dfrac{(2{,}390 \text{ lb/day})(94.2\%)}{(\text{Filter operation, hr/day})(100\%)}}{218 \text{ ft}^2}$$

Think of the above equation as:

$$1.93 \text{ lb/hr/ft}^2 = \dfrac{\dfrac{(2{,}390 \text{ lb/day})(94.2\%)}{(\text{Filter operation, hr/day})(100\%)}}{\dfrac{218 \text{ ft}^2}{1}}$$

The mathematical rule states you invert and multiply, as follows:

$$1.93 \text{ lb/hr/ft}^2 = \dfrac{(2{,}390 \text{ lb/day})(94.2\%)}{(\text{Filter operation, hr/day})(100\%)} \times \dfrac{(1)}{218 \text{ ft}^2}$$

Simplify:

$$1.93 \text{ lb/hr/ft}^2 = \dfrac{(2{,}390 \text{ lb/day})(1)(94.2\%)}{(\text{Filter operation, hr/day})(218 \text{ ft}^2)(100\%)}$$

Rearrange the equation to solve for the unknown, filter operation time.

$$\text{Filter operation, hr/day} = \dfrac{(2{,}390 \text{ lb/day})(1)(94.2\%)}{(1.93 \text{ lb/hr/ft}^2)(218 \text{ ft}^2)(100\%)} = \mathbf{5.35 \text{ hr/day}}$$

216. Given the following data, calculate the time in hr/day for a vacuum filter to process 5,880 lb/day:

Vacuum filter surface area = 249 ft²
Solids recovery averages 93.5%
Percent yield averages 1.87 lb/hr/ft²

Equation: **Filter yield, lb/hr/ft²** $= \dfrac{\text{(Solids, lb/day)(Percent recovery)}}{\dfrac{\text{(Filter operation, lb/day)(100\%)}}{\text{Filter area ft}^2}}$

Substitute values and solve.

$1.87 \text{ lb/hr/ft}^2 = \dfrac{(5,880 \text{ lb/day})(93.5\%)}{\dfrac{(\text{Filter operation, hr/day})(100\%)}{249 \text{ ft}^2}}$

Simplify:

$1.87 \text{ lb/hr/ft}^2 = \dfrac{(5,880 \text{ lb/day})(1)(93.5\%)}{(\text{Filter operation, hr/day})(249 \text{ ft}^2)(100\%)}$

Rearrange the equation to solve for the unknown, filter operation time.

$\text{Filter operation, hr/day} = \dfrac{(5,880 \text{ lb/day})(1)(93.5\%)}{(1.87 \text{ lb/hr/ft}^2)(249 \text{ ft}^2)(100\%)} = \textbf{11.8 hr/day}$

BIOSOLIDS VOLUME INDEX AND BIOSOLIDS DENSITY INDEX CALCULATIONS

Both of these calculations help determine the pumping rate of return biosolids and are a measure of the settling characteristics of the activated biosolids. They also exhibit how well the liquids/solids separation in the activated biosolids system is performing its role on the biological floc that has been generated. The goal is to produce a small volume of biosolids and thus reduce the pumping rate that is required for the solids to stay in circulation.

217. The mixed liquor suspended solids (MLSS) for an aeration tank is 2,810 mg/L. What is the biosolids volume index (BVI), if during a 30-minute settleability test 213 mL of biosolids settled in a 1-liter graduated cylinder?

Equation: **BVI** $= \dfrac{\text{(Settled biosolids, mL/L)}(1,000 \text{ mg/g})}{\text{MLSS, mg/L}}$

Substitute values and solve.

$\text{BVI} = \dfrac{(213 \text{ mL/L})(1,000 \text{ mg/g})}{2,810 \text{ mg/L}} = 75.8 \text{ mL/g, which reduces to } \textbf{75.8 BVI}$

218. **What is the biosolids density index (BDI) for an aeration tank, if the concentration of mixed liquor suspended solids (MLSS) is 2,385 mg/L and the settleability test result shows that 273 mL of activated biosolids settled in a 1-liter graduated cylinder?**

Equation: $\mathbf{BDI} = \dfrac{(\text{MLSS, mg/L})\,(100)}{(\text{Settled biosolids, mL/L})\,(1{,}000\text{ mg/g})}$

Substitute values and solve.

$\text{BDI} = \dfrac{(2{,}385\text{ mg/L})\,(100)}{(273\text{ mL/L})\,(1{,}000\text{ mg/g})} = 0.874$ g/mL, which reduces to **0.874 BDI**

SETTLEABLE SOLIDS CALCULATIONS

These tests are performed on samples from either the clarifier's influent or effluent or from a sedimentation tank. They are used to determine the percent and thus the efficiency of settleable solids. Calculations based on these tests follow:

219. **A 2,000.0 mL of activated sludge was collected in a graduated cylinder. What is the percent of settleable solids, if after exactly 30 minutes the sludge solids that settled totaled 381 mL?**

Equation: **Percent settleable solids** $= \dfrac{(\text{Settled sludge, mL})\,(100\%)}{\text{Sample size, mL}}$

Substitute values and solve.

Percent settleable solids $= \dfrac{(381\text{ mL})\,(100\%)}{2{,}000.0\text{ mL}} = 19.05\%$, round to **19.0% settled solids**

220. Given the following data, calculate the percent settleable solids:

Activated sludge sample = 0.750 gallon poured into a large graduated cylinder
Settling time is exactly = 30 minutes
Sludge solids in graduated cylinder = 509 mL

First, convert the sample in gallons to mL.

Number of mL = (0.750 gal)(3,785 mL/gal) = 2,838.75

Equation: **Percent settleable solids** $= \dfrac{(\text{Settled sludge, mL})(100\%)}{\text{Sample size, mL}}$

Substitute values and solve.

Percent settleable solids $= \dfrac{(509 \text{ mL})(100\%)}{2,838.75 \text{ mL}} = 17.93\%$, round to **17.9% Settled solids**

COMPOSTING CALCULATIONS

Composting is an aerobic biological process. This process decomposes organic matter to a stable end product. The optimum moisture content for composting ranges from 50 to 60% water. Several different composting calculations follow.

221. Given the following information, what is the percent moisture content of a composting blend?

18,880 lb of sludge was added and mixed with 9,275 lb of compost
Compost = 60.2% solids
Added sludge = 19.8% solids

Since percent moisture needs to be solved, first determine the percent moisture of both the compost and sludge.

Compost percent moisture = 100% − 60.2% solids = 39.8% moisture content

Sludge percent moisture = 100% − 19.8% solids = 80.2% moisture content

Equation: **Mixture's percent moisture** =

$$\frac{[(\text{Sludge, lb})(\text{Percent moisture}) + (\text{Compost, lb})(\text{Percent moisture})]\,100\%}{\text{Sludge, lb} + \text{Compost, lb}}$$

Substitute values and solve.

Mixture's percent moisture $= \dfrac{[(18{,}880\text{ lb})(80.2\%/100\%) + (9{,}275\text{ lb})(39.8\%/100\%)]\,100\%}{18{,}880\text{ lb} + 9{,}275\text{ lb}}$

Mixture's percent moisture $= \dfrac{(15{,}141.76\text{ lb} + 3{,}691.45\text{ lb})\,100\%}{28{,}155\text{ lb}} = \mathbf{66.9\%}$

222. **Given the following data for blending compost (BC) with wood chips, calculate the percent of the blended compost:**

Bulk density of sludge = 1,690 lb/yd³
Sludge volume = 15.3 yd³
Sludge solids content = 16.3%
Density of wood chips = 740 lb/yd³
Wood chip solids = 53.9%
Mix ratio (MR) of wood chips to sludge = 3.50 to 1

Equation: **Percent solids BC** =

$$\frac{[(\text{Sludge, yd}^3)(\text{lb/yd}^3)(\%\text{ solids, sludge}) + (\text{Sludge, yd}^3)(\text{MR})(\text{lb/yd}^3)(\%\text{ solids, chips})]\,(100\%)}{(\text{Sludge, yd}^3)(\text{lb/yd}^3) + (\text{Sludge, yd}^3)(\text{MR})(\text{lb/yd}^3)}$$

Substitute values and solve.

Percent solids BC =

$$\frac{[(15.3\text{ yd}^3)(1{,}690\text{ lb/yd}^3)(16.3\%/100\%) + (15.3\text{ yd}^3)(3.50)(740\text{ lb/yd}^3)(53.9\%/100\%)]\,(100\%)}{(15.3\text{ yd}^3)(1{,}690\text{ lb/yd}^3) + (15.3\text{ yd}^3)(3.50)(740\text{ lb/yd}^3)}$$

Percent solids BC $= \dfrac{(4{,}214.691 + 21{,}358.953)(100\%)}{25{,}857 + 39{,}627} = \dfrac{(25{,}573.644)(100\%)}{65{,}484}$

Percent solids BC = 39.05%, round to **39.0% Solids in blended compost**

223. Given the following parameters, calculate the amount of wet compost a composting site can process in lb/day and tons/day:

Compost site capacity = 9,950 yd³
Compost cycle averages = 20.1 days
Bulk density of compost averages 962 lb/yd³

Equation: **Compost cycle, days** = $\dfrac{(\text{Site capacity, yd}^3)(\text{Density of compost, lb/yd}^3)}{x \text{ lb/day}}$

$20.1 \text{ days} = \dfrac{(9{,}950 \text{ yd}^3)(962 \text{ lb/yd}^3)}{x \text{ lb/day}}$

Solve for x.

$x \text{ lb/day} = \dfrac{(9{,}950 \text{ yd}^3)(962 \text{ lb/yd}^3)}{(20.1 \text{ days})}$

$x \text{ lb/day} = 476{,}213.93 \text{ lb/day}$, round to **476,000 lb/day Capacity**

Lastly, calculate the capacity in tons.

$\text{Site capacity, tons} = \dfrac{476{,}213.93 \text{ lb/day}}{2{,}000 \text{ lb/ton}} = 238.107 \text{ tons}$, round to **238 tons**

CHEMISTRY AND LABORATORY PROBLEMS

Operators should have a thorough understanding of many laboratory calculations, for they help in evaluating plant processes and efficiencies. Following are a few examples:

224. Given the following data, what is the number of grams (g) in 1 mole of potassium permanganate (KMnO₄)?

Potassium (K) = 39.102 g/mole
Manganese = 54.938 g/mole
Oxygen = 15.999 g/mole

Equation: **Grams/mole = K g/mole + Mn g/mole + 4(O g/mole)**

Grams/mole = 39.102 g/mole + 54.938 g/mole + 4(15.999 g/mole)

Grams/mole = 39.102 g/mole + 54.938 g/mole + 63.996 g/mole = **158.036 g/mole for KMnO₄**

225. If 125 grams (g) of nitric acid (HNO_3) are dissolved in 1 liter of solution, how many moles of HNO_3 were used given that 1 mole equals 63.01 grams?

Equation: $\textbf{Moles} = \dfrac{\text{Grams of chemical}}{\text{Gram formula weight}}$

$\text{Moles} = \dfrac{125\,\text{g}}{63.01\,\text{g/mole}} = 1.9838$ moles, round to **1.98 moles of HNO_3**

226. What is the normality (N) of a NaOH solution if 1.85 equivalents are dissolved in 2.75 liters of solution?

Equation: $\textbf{Normality} = \dfrac{\text{Number of equivalents of solute}}{\text{Liters of solution}}$

$\text{Normality} = \dfrac{1.85\,\text{equivalents}}{2.75\,\text{liters}} = \textbf{0.673 N NaOH}$

227. Calculate the unseeded BOD_5 in mg/L, given the following data:

Start of test bottle dissolved oxygen (DO) = 7.4 mg/L
Bottle was incubated for 5 days in the dark at 20°C
After 5 days DO = 2.3 mg/L
Sample size = 130 mL
Total volume = 300 mL (assume two significant figures)

Equation: $\textbf{BOD}_5$ **unseeded, mg/L =**
$\dfrac{(\text{Initial DO, mg/L} - \text{Final DO, mg/L})(\text{Total volume, mL})}{\text{Sample volume, mL}}$

$BOD_5 \text{ unseeded, mg/L} = \dfrac{(7.4\,\text{mg/L} - 2.3\,\text{mg/L})(300\,\text{mL})}{130\,\text{mL}} = 11.769$ mg/L, round to **12 mg/L BOD_5**

228. Calculate the seeded BOD_5 in mg/L, given the following data:

Sample size = 125 mL
Initial DO = 8.3 mg/L
Final DO = 2.7 mg/L
BOD_5 of seed stock = 86 mg/L
Seed stock = 3.5 mL
Total volume = 300 mL (assume two significant figures)

First, calculate the seed correction in mg/L.

Equation: **Seed correction, mg/L** $= \dfrac{(\text{BOD}_5 \text{ of seed stock, mg/L})(\text{Seed stock, mg/L})}{\text{Total volume, mL}}$

Seed correction, mg/L $= \dfrac{(86 \text{ mg/L})(3.5 \text{ mg/L})}{300 \text{ mL}} = 1.00 \text{ mg/L}$

Next, calculate the BOD_5 seeded in mg/L.

Equation:

BOD_5 seeded, mg/L $=$

$\dfrac{(\text{Initial DO, mg/L} - \text{Final DO, mg/L} - \text{Seed correction, mg/L})(\text{Total volume, mL})}{\text{Sample volume, mL}}$

BOD_5 seeded, mg/L $= \dfrac{(8.3 \text{ mg/L} - 2.7 \text{ mg/L} - 1.00 \text{ mg/L})(300 \text{ mL})}{125 \text{ mL}}$

BOD_5 seeded, mg/L $= \dfrac{(4.6 \text{ mg/L})(300 \text{ mL})}{125 \text{ mL}} = 11.04 \text{ mg/L, round to }$ **11 mg/L BOD_5**

BASIC ELECTRICITY PROBLEMS

Operators should have a basic understanding of electrical calculations, and they must always exercise safety in dealing with electricity at wastewater treatment plants or anywhere.

229. **What is the voltage (E) on a circuit, if the current is 5 amperes (I) and the resistance (R) is 15 ohms?**

Equation: **Voltage = (Amps)(Resistance, ohms)** *or* **E = IR**

Substitute values and solve.

Voltage = (5 amps)(15 ohms) = **75 volts**

230. **What is the resistance on a circuit if the voltage is 110 and the amperes is 25?**

Equation: **Resistance, ohms = Voltage/Amps**

Substitute values and solve.

Resistance, ohms = 110 Volts/25 Amps = **4.4 ohms**

231. **A digester tank has a level capacity of 24.5 ft. Currently there is 17.3 ft of sludge water in the tank. What would the SCADA reading be on the board in mA for a 4 mA to 20 mA signal?**

Equation: **Current process reading** $= \dfrac{(\text{Live signal, mA} - 4\text{ mA offset})(\text{Maximum capacity})}{16\text{ milliamp span}}$

Substitute values and solve.

$17.3\text{ ft} = \dfrac{(\text{Live signal, mA} - 4\text{ mA offset})(24.5\text{ ft})}{16\text{ mA span}}$

Rearrange equation to solve for the current number of milliamps.

$\text{Live signal, mA} - 4\text{ mA offset} = \dfrac{(17.3\text{ ft})(16\text{ mA})}{24.5\text{ ft}}$

$\text{Live signal, mA} = \dfrac{(17.3\text{ ft})(16\text{ mA})}{24.5\text{ ft}} + 4\text{ mA}$

Live signal, mA = 11.298 mA + 4 mA = 15.298 mA, round to **15.3 mA**

232. **The SCADA system at a water plant uses a 4 mA to 20 mA signal to monitor the speed of the chemical pumps (range 0 to 100%). If the readout on a SCADA board reads 13.0 mA, what is the percent speed of the chemical pump?**

Equation: **Current process reading** $= \dfrac{(\text{Live signal, mA} - 4\text{ mA offset})(\text{Maximum capacity})}{16\text{ milliamp span}}$

Substitute values and solve.

$$\text{Pump speed, percent} = \frac{(13\,\text{mA} - 4\,\text{mA offset})(100\%)}{16\,\text{mA span}}$$

$$\text{Pump speed, percent} = \frac{(9\,\text{mA})(100\%)}{16\,\text{mA span}} = 56.25\%, \text{round to } \mathbf{56\%}$$

KILOWATT DETERMINATIONS

As above, operators should have a basic understanding of kilowatt calculations, and they must always exercise safety in dealing with electricity at wastewater treatment plants or anywhere.

233. **If a 210-hp pump operates 8.5 hr/day, how many kW/day of energy will be consumed?**

Equation: **Kilowatt-hr/day = (hp)(0.746 kW/hp)(Operating time, hr/day)**

Kilowatt-hr/day = (210 hp)(0.746 kW/hp)(8.5 hr/day)

Kilowatt-hr/day = 1,331.6 kW-hr/day, round to **1,300 kW-hr/day**

234. **Given the following data, calculate the cost to run a pump in dollars and cents for one 30-day month:**

Pump's hp = 150
Operating time = 4.75 hr/day
Cost = $0.085/kW-hr

First, calculate the kilowatt-hours/day.

Kilowatt-hr/day = (hp)(0.746 kW/hp)(Operating time, hr/day)

Kilowatt-hr/day = (150 hp)(0.746 kW/hp)(4.75 hr/day) = 531.525 kW-hr/day

Next, calculate the cost.

Cost = (531.525 kW-hr/day)($0.085/kW-hr)(30 days/month)

Cost = **$1,355.39/month**

1. A chemical pump discharges 217 mL of alum at a speed setting of 52% and a stroke setting of 35%. If the alum pump's speed is increased to 58% and the stroke setting remains the same, what should be the mL output from the pump? Assume pump has a linear output.

2. What is the percent volatile matter (VM) reduction for a digester, if the raw biosolids VM is 64.3% and the VM digested biosolids is 48.1%?

3. Given the following data, how many lb/day of volatile solids (VS) are pumped to a digester?

    Pumping rate = 4.7 gpm
    Solids content = 4.04%
    Volatile solids = 59.1%
    Specific gravity of sludge = 1.03

4. A solution containing 840 gallons of 9.8% hypochlorite is required. How many gallons of a 12.5% solution must be mixed with a 3.8% solution to make the required solution? Assume three significant figures. Solve the problem using the dilution triangle.

5. What is the weir overflow rate in gpd/ft, if the flow is 0.377 mgd and the radius of the clarifier is 35.1 feet?

6. A 36-inch sewage pipeline is flowing at a velocity of 1.41 ft/s and the depth of the sewage averages 8.25 inches. Determine the flow in the pipeline in gpm.

7. Given the following data, determine the rate a pump discharges into a tank in gpm:

    Duration pump operates = 23 hr and 18 minutes
    Tank diameter = 36.7 ft
    Wastewater level at beginning of pumping = 8.56 ft
    Wastewater level at end of pumping = 15.92 ft

8. Calculate the solids loading rate on a secondary clarifier with a diameter of 52.0 ft, flow rate of 1,120,000 gpd, and a mixed liquor suspended solids (MLSS) of 3,570 mg/L.

9. Given the following data, calculate the lb/day (d) of solids pumped to a sludge thickener:

    Sludge sample = 1,805.519 grams (g)
    Solids content after drying = 82.446 g
    Pump operates exactly 10 minutes every 1.0 hours
    Pump rate = 19.5 gpm
    Specific gravity (sp gr) = 1.03
    Clarifier effluent flow = 1.36 mgd

10. Given the following data, calculate the amount of solids and volatile solids removed in lb/day:

Pumping rate = 192 gpm
Pump frequency = 24 times/day
Pumping cycle = 8 minutes exactly per cycle
Solids = 3.24%
Volatile solids (VS) = 62.5%

11. Find the motor horsepower (mhp) for a pump with the following parameters:

Motor efficiency (ME): 89.2%
Total head (TH): 239 ft
Pump efficiency (PE): 77.9%
Flow: 1.65 mgd

12. How many gpd of a 11.5% sodium hypochlorite solution are needed to disinfect a flow of 1,830,000 gallons, if the dosage required is 9.5 mg/L and the specific gravity of the hypochlorite is 1.02?

13. How much sulfur dioxide in lb/day needs to be applied to dechlorinate a wastewater, if the flow is 2,360,000 gpd, the chlorine residual is 1.43 mg/L, and the sulfur dioxide must be 3.0 mg/L higher than the chlorine residual?

14. A wastewater treatment plant is treating 1,460 gpm with a polymer solution that has a specific gravity (sp gr) of 1.38. If the chemical pump delivers 184 mL in exactly 5 minutes, what is the polymer dosage in mg/L?

15. Given the following data, calculate the feed rate of a polymer solution in mL/min:

Influent flow = 1,025 gpm
Polymer dose = 8.36 mg/L
Polymer solution specific gravity = 1.28
Polymer percent purity = 33.0%

16. Determine the feed rate of dry alum in lb/day, if the drawdown in exactly 10 minutes was 172.95 grams (g) and the flow is 1,875,000 gpd.

17. What is the organic loading rate for a trickling filter that is 68.4 ft in diameter and 5.1 ft deep in lb BOD_5/d/1,000 ft^3, if the primary effluent flow is 2.94 mgd and the BOD_5 is 177 mg/L?

18. What is the loading on a digester in lb volatile solids (VS)/day/ft^3, if the sludge flow into the digester is 27,480 gpd, the digester is 59.9 ft in diameter, the sludge level is 18.7 ft, and the sludge is 4.88% solids with a specific gravity of 1.03 and 68.7% volatile solids?

19. Given the following data, calculate the mean cell residence time (MCRT):

Flow = 1.76 mgd
Aeration tank volume = 547,000 gallons
Clarifier tank volume = 285,000 gallons
MLSS = 2,780 mg/L
Waste rate = 19,350 gpd
Waste activated sludge (WAS) = 6,970 mg/L
Effluent TSS = 16.5 mg/L

20. How many pounds of mixed liquor suspended solids (MLSS) are being aerated, if the aeration tank is 55.3 ft in diameter, with a sludge height of 18.6 ft, the concentration of MLSS is 2,220 mg/L, and the specific gravity of the MLSS is 1.03?

21. Given the following data, determine the BOD$_5$ removal in lb/day from a trickling filter.

Plant influent flow = 1.44 mgd
Influent BOD$_5$ concentration = 261 mg/L
Effluent BOD$_5$ concentration = 139 mg/L

22. What is the food-to-microorganism (F/M) ratio for an aeration tank that is 51.4 ft in diameter, with a liquid level of 15.8 ft, if the primary effluent flow averages 2,860,000 gpd, the MLVSS is 2,870 mg/L, and the BOD$_5$ is 296 mg/L?

23. Given the following data, determine the seed sludge required in gallons:

Digester has a radius of 25.2 ft
Liquid level in digester is 21.3 ft
Requires 18.0% seed sludge

24. Given the following data, what is the feed time for a basket centrifuge thickener?

Basket centrifuge thickener capacity = 24 ft^3
Sludge flow rate = 48,800 gpd
Solids concentration = 6,650 mg/L
Percent solids = 6.35%

25. Given the following data for blending compost (BC) with wood chips, calculate the percent of the blended compost:

Bulk density of sludge = 1,685 lb/yd^3
Sludge volume = 17.4 yd^3
Sludge solids content = 20.6%
Density of wood chips = 640 lb/yd^3
Wood chip solids = 54.2%
Mix ratio (MR) of wood chips to sludge = exactly 3 to 1

1. **A chemical pump discharges 217 mL of alum at a speed setting of 52% and a stroke setting of 35%. If the alum pump's speed is increased to 58% and the stroke setting remains the same, what should be the mL output from the pump? Assume pump has a linear output.**

This problem can be solved using a ratio, as follows:

$$\frac{\text{Alum dosage}_1, \text{mL}}{\text{Speed setting}_1, \%} = \frac{\text{Alum dosage}_2, \text{mL}}{\text{Speed setting}_2, \%}$$

Substitute values and solve.

$$\frac{217 \text{ mL}}{52\%} = \frac{x \text{ mL}}{58\%}$$

$$x \text{ mL} = \frac{(58\%)(217 \text{ mL})}{52\%} = 242.04 \text{ mL, round to } \mathbf{240 \text{ mL}}$$

2. **What is the percent volatile matter (VM) reduction for a digester, if the raw bio-solids VM is 64.3% and the VM digested biosolids is 48.1%?**

Equation: $\dfrac{(\text{Percent influent VM} - \text{Percent effluent VM})\,(100\%)}{[\text{Percent influent VM} - (\text{Percent influent VM})\,(\text{Percent effluent VM})]}$

Percent VM reduction =

First, convert percentages to decimal form for easier substitution.

Raw biosolids = 64.3%/100% = 0.643

Digested VM = 48.1%/100% = 0.481

Substitute values and solve.

Percent reduction = $\dfrac{(0.643 - 0.481)\,(100\%)}{[0.643 - (0.643)\,(0.481)]}$

Simplify:

Percent reduction = $\dfrac{(0.162)\,(100\%)}{(0.643 - 0.309283)}$

Percent reduction = $\dfrac{(0.162)\,(100\%)}{0.333717}$ = **48.5% VM reduction**

3. **Given the following data, how many lb/day of volatile solids (VS) are pumped to a digester?**

Pumping rate = 4.7 gpm
Solids content = 4.04%
Volatile solids = 59.1%
Specific gravity of sludge = 1.03

First, determine the lb/gal for the sludge.

Sludge, lb/gal = (8.34 lb/gal)(1.03) = 8.59 lb/gal

Next, convert gpm to gpd.

Number of gpd = (4.7 gpm)(1,440 min/day) = 6,768 gpd

Equation:

VS, lb/day =
(Number of gpd to digester) $\dfrac{\text{(Percent solids)}}{100\%}\dfrac{\text{(Percent VS)}}{100\%}$ **(8.34 lb/gal or sludge, lb/gal)**

VS, lb/day = (6,768 gpd Solids) $\dfrac{(4.04\%)}{100\%}\dfrac{(59.1\% \text{ VS})}{100\%}$ (8.59 lb/gal)

VS, lb/day = 1,388 lb/day, round to **1,400 lb/day VS**

4. **A solution containing 840 gallons of 9.8% hypochlorite is required. How many gallons of a 12.5% solution must be mixed with a 3.8% solution to make the required solution? Assume three significant figures. Solve the problem using the dilution triangle.**

12.5%		6.0	6.0 parts of the 12.5% solution are required for every 8.7 parts.
	9.8%		
3.8%		2.7	2.7 parts of the 3.8% solution are required for every 8.7 parts.
		8.7	total parts

$\dfrac{6.0\,\text{parts}\,(840\,\text{gal})}{8.7\,\text{parts}}$ = **579 gallons of the 9.8% solution**

$\dfrac{2.7\,\text{parts}\,(840\,\text{gal})}{8.7\,\text{parts}}$ = **261 gallons of the 2.5% solution**
840 gallons

To make the 840 gallons of the 9.8% solution, mix 579 gallons of the12.5% solution with 261 gallons of the 3.8% solution.

5. **What is the weir overflow rate in gpd/ft, if the flow is 0.377 mgd and the radius of the clarifier is 35.1 feet?**

First, convert mgd to gpd.

Number of gpd = (0.377 mgd)(1,000,000 gal/mil) = 377,000 gpd

Next, calculate the length of the weir.

Weir length, ft = 2π(radius, ft)

Weir length, ft = 2(3.14)(35.1 ft) = 220.428 ft

Next, solve for the weir overflow rate.

Equation: **Weir overflow rate, gpd/ft** $= \dfrac{\text{Flow, gpd}}{\text{Weir length, ft}}$

Weir overflow rate, gpd/ft $= \dfrac{377,000 \text{ gpd}}{220.428 \text{ ft}} = 1{,}710.3$ gpd/ft, round to **1,710 gpd/ft**

6. **A 36-inch sewage pipeline is flowing at a velocity of 1.41 ft/s and the depth of the sewage averages 8.25 inches. Determine the flow in the pipeline in gpm.**

First, divide the depth of sewage flow by the diameter of the pipe. Converting inches to feet is not necessary in this step.

Ratio = depth/Diameter = 8.25 in./36 in. = 0.2292, round to 0.23

Next, determine the factor that needs to be used.

In Appendix D, look up 0.23 under the column d/D. The number immediately to the right will be the factor that needs to be used. In this case it is 0.1365. This will be the number used rather than 0.785.

Next, convert the pipe's diameter from inches to feet.

Number of feet $= \dfrac{36 \text{ in.}}{12 \text{ in./ft}} = 3.0$

Equation: **Flow, ft³/sec = (Area, ft²)(Velocity, ft/s)**

Where the area = (Factor)(Diameter)^2

Substitute values and solve.

Flow, ft^3/s = (0.1365)(3.0 ft)(3.0 ft)(1.41 ft/s) = 1.732 ft^3/s

Now, convert ft^3/s to gpm.

Flow, gpm = (1.732 ft^3/s)(60 s/min)(7.48 gal/ft^3) = 777.32 gpm, round to **780 gpm**

7. **Given the following data, determine the rate a pump discharges into a tank in gpm:**

Duration pump operates = 23 hr and 18 minutes
Tank diameter = 36.7 ft
Wastewater level at beginning of pumping = 8.56 ft
Wastewater level at end of pumping = 15.92 ft

First, find the number of minutes the pump worked.

Number of min = (23 hr)(60 min/hr) + 18 min = 1,398 min

Next, calculate the change in level during pumping.

Level change, ft = 15.92 ft − 8.56 ft = 7.36 ft

Next, calculate the volume in gallons added to the tank by the pump.

Equation: Volume, gal = (0.785)(Diameter)^2(Level change, ft)(7.48 gal/ft^3)

Volume, gal = (0.785)(36.7 ft)(36.7 ft)(7.36 ft)(7.48 gal/ft^3) = 58,207.80 gal

Now, calculate the pump's discharge rate in gpm.

Equation: **Pump's discharge rate, gpm** $= \dfrac{\text{Discharge, gal}}{\text{Time, min}}$

Substitute values and solve.

Pump's discharge rate, gpm $= \dfrac{58,207.80 \text{ gal}}{1,398 \text{ min}} = 41.636$ gpm, round to **42 gpm**

8. **Calculate the solids loading rate on a secondary clarifier with a diameter of 52.0 ft, flow rate of 1,120,000 gpd, and a mixed liquor suspended solids (MLSS) of 3,570 mg/L.**

First, determine the area of the clarifier.

Area $= \pi r^2$ where r $=$ Diameter/2 $=$ 52.0 ft/2 $=$ 26.0 ft

Area $=$ (3.14)(26.0 ft)^2 $=$ (3.14)(26.0 ft)(26.0 ft) $=$ 2,122.64 ft^2

Next, convert gpd to mgd.

Number of mgd $=$ (1,120,000 gpd) / (1,000,000 gal/M) $=$ 1.12 mgd

Finally, calculate the solids loading rate.

Equation: **Solids loading rate** $= \dfrac{(\text{MLSS, mg/L}) (\text{mgd}) (8.34 \text{ lb/gal})}{\text{Area, ft}^2}$

Solids loading rate $= \dfrac{(3,570 \text{ mg/L}) (1.12 \text{ mgd}) (8.34 \text{ lb/gal})}{2,122.64 \text{ ft}^2}$

Solids loading rate $=$ 15.71 lb of solids/d/ft^2, round to **15.7 lb of solids/d/ft^2**

9. Given the following data, calculate the lb/day (d) of solids pumped to a sludge thickener:

Sludge sample = 1,805.519 grams (g)
Solids content after drying = 82.446 g
Pump operates exactly 10 minutes every 1.0 hours
Pump rate = 19.5 gpm
Specific gravity (sp gr) = 1.03
Clarifier effluent flow = 1.36 mgd

First, determine the percent solids in the sludge.

Equation: **Solids, % = (Dry solids, g/Sludge sample, g)(100%)**

Solids, % = (82.446 g)(100%)/1,805.519 g = 4.5663% solids

Now, calculate the solids pumped in lb/day.

Equation:

Solids, lb/day = (Pumping, min/d)(24 hr/d)(Pump rate, gpm)(8.34 lb/gal)(sp gr of sludge)

Solids, lb/day = (10 min/hr)(24 hr/day)(19.5 gpm)(8.34 lb/gal)(1.03 sp gr)(4.5663%/100 %)

Solids, lb/day = 1,835.75 lb/day, round to **1,840 lb/day of solids**

10. **Given the following data, calculate the amount of solids and volatile solids removed in lb/day:**

Pumping rate = 192 gpm
Pump frequency = 24 times/day
Pumping cycle = 8 minutes exactly per cycle
Solids = 3.24%
Volatile solids (VS) = 62.5%

First, determine the solids removal in lb/day.

Equation: **Solids, lb/day =**

(Time, min/cycle)(cycles/day)(Pump rate, gpm)(8.34 lb/gal)(Percent solids)

Substitute values and solve.

Solids, lb/day = (8 min/cycle)(24 cycles/day)(192 gpm)(8.34 lb/gal)(3.24%/100%)

Solids, lb/day = 9,961.24 lb/day, round to **9,960 lb/day Solids**

Next, calculate the amount of volatile solids removed in lb/day.

Equation: **VS, lb/day =**

(Time, min/cycle)(cycles/day)(Pump rate, gpm)(8.34 lb/gal)(Percent, solids)(Percent VS)

Substitute values and solve.

VS, lb/day = (8 min/cycle)(24 cycles/day)(192 gpm)(8.34 lb/gal)(3.24%/100%)(62.5%/100%)

VS, lb/day = 6,225.78 lb/day, round to **6,230 lb/day VS**

11. **Find the motor horsepower (mhp) for a pump with the following parameters:**

Motor efficiency (ME): 89.2%
Total head (TH): 239 ft
Pump efficiency (PE): 77.9%
Flow: 1.65 mgd

First, convert mgd to gpm.

Gpm = (1.65 mgd)(1,000,000/mil)(1 day/1,440 min) = 1,145.83 gpm

The equation for determining the mhp with the given data is different then the problem above.

Equation: $\mathbf{mhp} = \dfrac{(\text{Flow, gpm})(\text{TH, ft})}{(3,960)(\text{ME})(\text{PE})}$

Substitute values and solve.

$$mhp = \frac{(1,145.83 \text{ gpm})(239 \text{ ft})}{(3,960)(89.2\%/100\% \text{ ME})(77.9\%/100\% \text{ PE})}$$

mhp = 99.52 mhp, round to **99.5 mhp**

12. **How many gpd of a 11.5% sodium hypochlorite solution are needed to disinfect a flow of 1,830,000 gallons, if the dosage required is 9.5 mg/L and the specific gravity of the hypochlorite is 1.02?**

First, convert gpd to mgd.

Number of mgd = $\dfrac{1,830,000 \text{ gpd}}{1,000,000/\text{mil}}$ = 1.83 mgd

Next, determine the lb/gal for the hypochlorite solution.

Hypochlorite, lb/gal = (8.34 lb/gal)(1.02 sp gr) = 8.5068

Next, using the "pounds equation," calculate the lb/day of chlorine needed.

Equation: **Chlorine, lb/day = (Dosage, mg/L)(mgd)(8.34 lb/gal)**

Chlorine, lb/day = (9.5 mg/L)(1.83 mgd)(8.34 lb/gal) = 144.99 lb/day

Since the solution is not 100%, divide the percent hypochlorite into the lb/day of chlorine needed.

$$\text{Hypochlorite, lb/day} = \frac{144.99 \, \text{lb/day}}{11.5\%/100\%} = 1{,}260.78 \, \text{lb/day hypochlorite}$$

Lastly, determine the gpd of hypochlorite solution needed.

$$\text{Hypochlorite, gpd} = \frac{1{,}260.78 \, \text{lb/day}}{8.5068 \, \text{lb/gal}} = 148.2 \, \text{gpd, round to } \textbf{150 gpd sodium hypochlorite}$$

13. **How much sulfur dioxide (SO_2) in lb/day needs to be applied to dechlorinate a wastewater, if the flow is 2,360,000 gpd, the chlorine residual is 1.43 mg/L, the chlorine demand is 6.14 mg/L, and the sulfur dioxide must be 3.0 mg/L higher than the chlorine residual?**

First, determine how many mg/L of sulfur dioxide must be applied. This is the chlorine residual plus the amount that is required higher than the chlorine residual.

SO_2, mg/L = 1.43 mg/L + 6.14 mg/L + 3.0 mg/L = 10.57 mg/L SO_2

Next, convert gpd to mgd.

Number of mgd = (2,360,000 gpd) / (1,000,000/mil) = 2.36 mgd

Next, determine the number of lb/day of SO_2 needed.

Equation: **Number of lb/day SO_2 = (SO_2, mg/L)(Number of mgd)(8.34 lb/gal)**

Substitute values and solve.

Number of lb/day SO_2 = (10.57 mg/L)(2.36 mgd)(8.34 lb/gal)

Number of lb/day SO_2 = 208.04 lb/day, round to **210 lb/day SO_2**

14. **A wastewater treatment plant is treating 1,460 gpm with a polymer solution that has a specific gravity (sp gr) of 1.38. If the chemical pump delivers 184 mL in exactly 5 minutes, what is the polymer dosage in mg/L?**

First, determine the number of mL/min the pump is feeding.

Polymer feed, mg/L $= 184$ mL/5 min $= 36.8$ mL/min

Next, determine the lb/gal for the polymer.

Polymer, lb/gal $= (8.34$ lb/gal$)(1.38$ sp gr$) = 11.5092$ lb/gal

Next, convert gpm to mgd.

$$\text{Number of mgd} = \frac{(1,460 \text{ gpm})(1,440 \text{ min/day})}{1,000,000/\text{mil}} = 2.1024$$

Now, calculate the dosage using the following equation:

$$\textbf{Polymer dosage, mg/L} = \frac{(\text{mL/min})(1,440 \text{ min/day})(\text{Polymer, lb/gal})}{(3,785 \text{ mL/gal})(\text{mgd})(8.34 \text{ lb/gal})}$$

Substitute values and solve.

$$\text{Polymer dosage, mg/L} = \frac{(36.8 \text{ mL/min})(1,440 \text{ min/day})(11.5092 \text{ lb/gal})}{(3,785 \text{ mL/gal})(2.1024 \text{ mgd})(8.34 \text{ lb/gal})}$$

Polymer dosage, mg/L $= \textbf{9.19 mg/L}$

15. **Given the following data, calculate the feed rate of a polymer solution in mL/min:**

Influent flow = 1,025 gpm
Polymer dose = 8.36 mg/L
Polymer solution specific gravity = 1.28
Polymer percent purity = 33.0%

First, convert gpm to mgd.

$$\text{Number of mgd} = \frac{(1,025 \text{ gpm})(1,440 \text{ min/day})}{1,000,000/\text{mil}} = 1.476 \text{ mgd}$$

Next, determine the lb/gal for the polymer.

Polymer, lb/gal = (8.34 lb/gal)(1.28 sp gr) = 10.6752 lb/gal

Now, calculate the dosage using the following equation:

$$\textbf{Polymer dosage, mg/L} = \frac{(\text{mL/min})(1,440 \text{ min/day})(\text{Polymer, lb/gal})}{(3,785 \text{ mL/gal})(\text{mgd})(8.34 \text{ lb/gal})(\text{Percent Polymer})}$$

Rearrange the formula to solve for mL/min.

Polymer feed, mL/min =

$$\frac{(\text{Polymer dosage, mg/L})(3,785 \text{ mL/gal})(\text{mgd})(8.34 \text{ lb/gal})(\text{Percent Polymer})}{(1,440 \text{ min/day})(\text{Polymer, lb/gal})}$$

Substitute values and solve.

$$\text{Polymer feed mL/min} = \frac{(8.36 \text{ mg/L})(3,785 \text{ mL/gal})(1.476 \text{ mgd})(8.34 \text{ lb/gal})(33.0\%/100\%)}{(1,440 \text{ min/day})(10.6752 \text{ lb/gal})}$$

Polymer feed mL/min = **8.36 mL/min**

16. **Determine the feed rate of dry alum in lb/day, if the drawdown in exactly 10 minutes was 172.95 grams (g) and the flow is 1,875,000 gpd.**

First, determine the number of grams used per minute.

Alum, g = 172.95 g/10 min = 17.295 g/min

Equation: **Alum, lb/day** $= \dfrac{(\text{Number of g/min})\,(1{,}440 \text{ min/day})}{454 \text{ g/lb}}$

Substitute values and solve.

Alum, lb/day $= \dfrac{(17.295 \text{ g/min})\,(1{,}440 \text{ min/day})}{454 \text{ g/lb}} = 54.856$ lb/day, round to **54.9 lb/day Alum**

17. **What is the organic loading rate for a trickling filter that is 68.4 ft in diameter and 5.1 ft deep in lb BOD$_5$/d/1,000 ft^3, if the primary effluent flow is 2.94 mgd and the BOD$_5$ is 177 mg/L?**

First, determine the volume of the tricking filter in ft^3.

Volume, ft^3 = (0.785)(Diameter)^2(Depth, ft)

Volume, ft^3 = (0.785)(68.4 ft)(68.4 ft)(5.1 ft) = 18,730.61 ft^3

Next, factor out 1,000 ft^3 from the volume = (18.73061)(1,000 ft^3)

Next, determine the pounds of BOD$_5$/d/1,000 ft^3 using a modified version of the "pounds" equation.

Organic loading rate, lb BOD$_5$/d/1,000 ft^3 $= \dfrac{(\text{BOD}_5,\ \text{mg/L})\,(\text{Flow, mgd})\,(8.34 \text{ lb/gal})}{\text{Volume of trickling filter, ft}^3/1{,}000 \text{ ft}^3}$

Organic loading rate, lb BOD$_5$/d/1,000 ft^3 $= \dfrac{(177 \text{ mg/L BOD}_5)\,(2.94 \text{ mgd})\,(8.34 \text{ lb/gal})}{(18.73061)\,(1{,}000 \text{ ft}^3)}$

Organic loading rate, lb BOD$_5$/d/1,000 ft^3 = 231.7 lb BOD$_5$/d/1,000 ft^3 round to

Organic loading rate, lb BOD$_5$/d/1,000 ft^3 = **230 lb BOD$_5$/d/1,000 ft^3**

18. **What is the loading on a digester in lb volatile solids (VS)/day/ft³, if the sludge flow into the digester is 27,480 gpd, the digester is 59.9 ft in diameter, the sludge level is 18.7 ft, and the sludge is 4.88% solids with a specific gravity of 1.03 and 68.7% volatile solids?**

Equation: **Digester loading, lb VS/day/ft³** =

$$\frac{(Flow, gpd)(8.34\ lb/gal)(sp\ gr)(Percent\ sludge)(Percent\ volatile\ solids)}{(0.785)(Diameter)^2(Sludge\ level)}$$

Substitute values and solve.

$$\text{Digester loading, lb VS/day/ft}^3 = \frac{(27,480\ gpd)(8.34\ lb/gal)(1.03)(4.88\%/100\%)(68.7\%/100\%)}{(0.785)(59.9\ ft)(59.9\ ft)(18.7\ ft)}$$

$$\text{Digester loading, lb VS/day/ft}^3 = \frac{7,914\ lb\ VS/day}{52,670.19\ ft^3}$$

$$\text{Digester loading, lb VS/day/ft}^3 = \textbf{0.150 lb VS/day/ft}^3$$

19. **Given the following data, calculate the mean cell residence time (MCRT):**

Flow = 1.76 mgd
Aeration tank volume = 547,000 gallons
Clarifier tank volume = 285,000 gallons
MLSS = 2,780 mg/L
Waste rate = 19,350 gpd
Waste activated sludge (WAS) = 6,970 mg/L
Effluent TSS = 16.5 mg/L

First, convert the volumes for the tanks to mil gal.

$$\text{Aeration tank, mgd} = \frac{547,000\ gal}{1,000,000/mil} = 0.547\ \text{mil gal}$$

$$\text{Clarifier tank, mgd} = \frac{285,000\ gal}{1,000,000/mil} = 0.285\ \text{mil gal}$$

Next, convert the waste rate from gpd to mgd.

$$\text{Waste rate, mgd} = \frac{19,350\ gal}{1,000,000/mil} = 0.01935\ \text{mgd}$$

Next, calculate the MCRT.

Equation: **MCRT, days** =

$$\frac{(MLSS, mg/L)(\text{Aeration tank mil gal} + \text{Clarifier tank mil gal})(8.34 \text{ lb/gal})}{(WAS, mg/L)(\text{Waste rate, mgd})(8.34 \text{ lb/gal}) + (TSS, mg/L)(\text{Flow, mgd})(8.34 \text{ lb/gal})}$$

MCRT, days =

$$\frac{(2,780 \text{ mg/L MLSS})(0.547 \text{ mil gal} + 0.285 \text{ mil gal})(8.34 \text{ lb/gal})}{(6,970 \text{ mg/L})(0.01935 \text{ mgd})(8.34 \text{ lb/gal}) + (16.5 \text{ mg/L TSS})(1.76 \text{ mgd})(8.34 \text{ lb/gal})}$$

$$\text{MCRT, days} = \frac{(2,780 \text{ mg/L MLSS})(0.832 \text{ mil gal})(8.34 \text{ lb/gal})}{1,124.81 \text{ lb/day} + 242.1936 \text{ lb/day}}$$

$$\text{MCRT, days} = \frac{19,290.09 \text{ lb}}{1,367 \text{ lb/day}} = 14.11 \text{ days, round to } \textbf{14.1 days}$$

20. **How many pounds of MLSS are being aerated, if the aeration tank is 55.3 ft in diameter, with a sludge height of 18.6 ft, the concentration of MLSS is 2,220 mg/L, and the specific gravity of the MLSS is 1.03?**

First, determine how many gallons are in the aeration tank.

Number of gallons = (0.785)(Diameter)^2(Height, ft)(7.48 gal/ft^3)

Number of gallons = (0.785)(55.3 ft)(55.3 ft)(18.6 ft)(7.48 gal/ft^3) = 333,991 gal

Next, convert gallons to mil gal.

$$\text{Number of mil gal} = \frac{333,991 \text{ gal}}{1,000,000/\text{mil}} = 0.333991 \text{ mil gal}$$

Next, determine the lb/gal for the MLSS.

Number of lb/gal, MLSS = (8.34 lb/gal)(1.03 sp gr) = 8.59 lb/gal

Next, determine the pounds of MLSS under aeration using a modified version of the "pounds" equation because the MLSS weighs more than water (8.34 lb/gal).

Equation: **Number of lb MLSS = (MLSS, mg/L)(Number of mil gal)(MLSS lb/gal)**

Substitute values and solve.

Number of lb MLSS = (2,220 mg/L MLSS)(0.333991 mil gal)(8.59 lb/gal)

Number of lb MLSS = 6,369.14 lb MLSS, round to **6,370 lb MLSS**

21. **Given the following data, determine the BOD$_5$ removal in lb/day from a trickling filter.**

Plant influent flow = 1.44 mgd
Influent BOD$_5$ concentration = 261 mg/L
Effluent BOD$_5$ concentration = 139 mg/L

First, determine the amount of BOD$_5$ removed in mg/L by subtracting the influent BOD$_5$ from the effluent BOD$_5$.

BOD$_5$ removed, mg/L = (Influent BOD$_5$, mg/L – Effluent BOD$_5$, mg/L)

BOD$_5$ removed, mg/L = 261 mg/L – 139 mg/L = 122 mg/L BOD$_5$ removed

Next, solve the amount of BOD$_5$ removed in lb/day by used the "pounds" formula.

Equation: **Number of lb/day BOD$_5$ removed = (BOD$_5$, mg/L)(Number of mgd)(8.34 lb/gal)**

Substitute values and solve.

Number of lb/day BOD$_5$ removed = (122 mg/L BOD$_5$)(1.44 mgd)(8.34 lb/gal)

Number of lb/day BOD$_5$ removed = 1,465.17 lb/day, round to **1,470 lb/day BOD$_5$ removed**

22. What is the food-to-microorganism (F/M) ratio for an aeration tank that is 51.4 ft in diameter, with a liquid level of 15.8 ft, if the primary effluent flow averages 2,860,000 gpd, the MLVSS is 2,870 mg/L, and the BOD_5 is 296 mg/L?

First, calculate the number of gallons in the aeration tank.

Number of gallons = (0.785)(Diameter)^2(Height, ft)(7.48 gal/ft^3)

Number of gallons = (0.785)(51.4 ft)(51.4 ft)(15.8 ft)(7.48 gal/ft^3) = 245,106 gal

Next, convert gallons to mil gal.

$$\text{Number of mil gal} = \frac{245,106 \text{ gpd}}{1,000,000/\text{mil}} = 0.245106 \text{ mil gal}$$

Next, convert the effluent flow in gpd to mgd.

Number of mgd = 2,860,000 gal/1,000,000/mil = 2.86 mgd

Next, write the equation:

$$\textbf{F/M} = \frac{(BOD_5, \text{mg/L})(\text{Flow, mgd})}{(\text{mg/L MLVSS})(\text{Volume of tank, mil gal})}$$

Substitute values and solve.

$$\text{F/M} = \frac{(296 \text{ mg/L } BOD_5)(2.86 \text{ mgd})}{(2,870 \text{ mg/L MLVSS})(0.245106 \text{ mil gal})} = \textbf{1.20 F/M ratio}$$

23. Given the following data, determine the seed sludge required in gallons:

Digester has a radius of 25.2 ft
Liquid level in digester is 21.3 ft
Requires 18.0% seed sludge

First, determine the number of gallons in the digester.

Volume, gal = πr^2(Depth, ft)(7.48 gal/ft³)

Volume, gal = (3.14)(25.2 ft)(25.2 ft)(21.3 ft)(7.48 gal/ft³) = 317,696 gal

Next, use the following equation.

$$\textbf{Seed sludge, gal} = \frac{\text{(Capacity of digester) (Percent seed sludge required)}}{100\%}$$

Seed sludge, gal = (317,696 gallons)$\dfrac{(18.0\%)}{100\%}$

Seed sludge, gal = 57,185 gal, round to **57,200 gal of seed sludge**

24. Given the following data, what is the feed time for a basket centrifuge thickener?

Basket centrifuge thickener capacity = 24 ft³
Sludge flow rate = 48,800 gpd
Solids concentration = 6,650 mg/L
Percent solids = 6.35%

Know: 1% = 10,000 ppm or mg/L

First, convert the solids concentration in mg/L to percent.

Percent solids = $\dfrac{6,650\,\text{mg/L}}{10,000\,\text{mg/L/1\%}}$ = 0.6650%

Next, convert gpd to gpm.

Number of gpm = $\dfrac{48,800\,\text{gpd}}{1,440\,\text{min/day}}$ = 33.89 gpm

Now, calculate the feed time in minutes.

Equation: **Feed time, min** $= \dfrac{(\text{Capacity, ft}^3)\,(\text{Solids, \%}/100\%)\,(7.48\ \text{gal/ft}^3)\,(8.34\ \text{lb/gal})}{(\text{Flow, gpm})\,(\text{Solids concentration, \%}/100\%)\,(8.34\ \text{lb/gal})}$

Simplify equation by canceling out the 8.34 lb/gal and the 100%:

Feed time, min $= \dfrac{(\text{Capacity, ft}^3)\,(\text{Solids, \%})\,(7.48\ \text{gal/ft}^3)}{(\text{Flow, gpm})\,(\text{Solids concentration, \%})}$

Feed time, min $= \dfrac{(24\ \text{ft}^3)\,(6.35\%)\,(7.48\ \text{gal/ft}^3)}{(33.89\ \text{gpm})\,(0.6650\%)} = 50.58$ min, round to **51 min**

25. **Given the following data for blending compost (BC) with wood chips, calculate the percent of the blended compost:**

Bulk density of sludge = 1,685 lb/yd³
Sludge volume = 17.4 yd³
Sludge solids content = 20.6%
Density of wood chips = 640 lb/yd³
Wood chip solids = 54.2%
Mix ratio (MR) of wood chips to sludge = exactly 3 to 1

Equation: **Percent solids BC =**

$\dfrac{(\text{Sludge, yd}^3)\,(\text{lb/yd}^3)\,(\%\ \text{solids, sludge}) + (\text{Sludge, yd}^3)\,(\text{MR})\,(\text{lb/yd}^3)\,(\%\ \text{solids, chips})\,(100\%)}{(\text{Sludge, yd}^3)\,(\text{lb/yd}^3) + (\text{Sludge, yd}^3)\,(\text{Mix ratio})\,(\text{lb/yd}^3)}$

Substitute values and solve.

Percent solids BC =

$\dfrac{\big[(17.4\ \text{yd}^3)\,(1,685\ \text{lb/yd}^3)\,(20.6\%/100\%) + (17.4\ \text{yd}^3)\,(3)\,(640\ \text{lb/yd}^3)\,(54.2\%/100\%)\big]\,(100\%)}{(17.4\ \text{yd}^3)\,(1,685\ \text{lb/yd}^3) + (17.4\ \text{yd}^3)\,(3)\,(640\ \text{lb/yd}^3)}$

Percent solids BC $= \dfrac{(6,039.714 + 18,107.136)\,(100\%)}{29,319 + 33,408} = \dfrac{(24,146.85)\,(100\%)}{62,727}$

Percent solids BC = 38.495%, round to **38% Solids in blended compost**

APPENDIXES

APPENDIX A

COMMON CONVERSION FACTORS

AREA

1 acre (ac) = 43,560 square feet (ft^2)
1 acre-ft = 43,560 cubic feet (ft^3)
2.4711 ac = 1 hectare
1 hectare = 0.4047 acre

1 hectare = 10,000 square meters
1 square mile = 640 acres

CONCENTRATION

1% solution = 1 part in 100 parts 1 ppm = 1 milligram per liter (mg/L)
1% solution = 10,000 parts per million (ppm) 1 grain per gal (gpg) = 17.12 ppm

DENSITY

Water has a density of 1 gram per mL (1g/mL) or 8.34 lb/gal or 62.4 lb/ft^3

FLOW

1 miner's inch = 1.5 ft^3/min
1 ft^3/s = 448.8 gal/min (gpm)

1 ft^3/s = 0.6463 million gallons per day (mgd)
1 mgd = 1.547 ft^3/s

LENGTH

1 inch = 2.54 centimeter (cm)
100 cm = 1 meter (m)
1 m = 39.37 inches
1 m = 3.281 feet (ft

1 yard = 0.9144 m
1,000 m = 1 kilometer (km)
1 km = 1.609 miles

POWER

1 horsepower (hp) = 0.746 kilowatts (kw)
1 kw = 1.341 hp)

PRESSURE

1 lb per sq in (psi) = 2.307 ft. of water
1 foot of water = 0.4335 psi
1 atmosphere (atm) = 14.7 psi

1 atmosphere = 29.92 inches of mercury
1 atm = 33.90 ft of water
1 atm = 760 mm of mercury

TEMPERATURE

Degrees Fahrenheit (°F) = (9 °F/5 °C)(°C) + 32°F
Degrees Celsius (°C) = (°F − 32 °F)(5 °C /9 °F)

VOLUME

2 pints = 1 quart
8 pints = 1 gallon (gal)
4 quarts = 1 gallon
1 quart = 32 fluid ounces
1 gallon = 128 fluid ounces
1 mL = 1 cubic centimeter
1 cubic foot = 7.48 gallons
1 acre-foot = 325,829 gal

1 gallon = 3.785 liters
1 liter = 1.0567 quarts
1 liter = 1,000 milliliters (mL)
3,785 ml = 1 gallon
1,000 liters = 1 cubic meter
1 cubic meter = 35.3 ft^3
1 MG = 3.07 acre-feet

WEIGHT

1 gram (g) = 1,000 milligrams
1,000 gm = 1 kilogram (kg)
1 lb = 454 g
1 lb = 7,000 grains (gr)

1 kg = 2.205 pounds (lb)
2,000 lb = 1 ton
1 mg/L = 1 part per million (ppm)
1 grain per gal (gpg) = 17.1 ppm

APPENDIX B

SUMMARY OF WASTEWATER TREATMENT EQUATIONS

AREA EQUATIONS

Area of a rectangle $= (\text{Length})(\text{Width})$

Area of a circle(tank) $= (0.785)(\text{Diameter})^2$ or πr^2

Area of a parallelogram $= (\text{Base})(\text{Height})$

Area of a trapezoid $= \dfrac{(\text{Altitude})(\text{Base}_1 + \text{Base}_2)}{2}$

BASIC ELECTRICITY FORMULAS

Voltage $= (\text{Amps})(\text{Resistance, ohms})$

Resistance, ohms $= \text{Voltage/Amps}$

Current process reading $= \dfrac{(\text{Live signal, mA} - 4\,\text{mA offset})(\text{Maximum capacity})}{16\,\text{milliamp span}}$

BIOCHEMICAL OXYGEN DEMAND LOADING EQUATIONS

$\text{BOD}_5, \text{lb/day} = (\text{BOD}_5, \text{mg/L})(\text{Number of mgd})(8.34 \text{ lb/gal})$

BIOCHEMICAL OXYGEN DEMAND UNSEEDED AND SEEDED EQUATIONS

$$\text{BOD}_5 \text{ unseeded, mg/L} = \frac{(\text{Initial DO, mg/L} - \text{Final DO, mg/L})(\text{Total volume, mL})}{\text{Sample volume, mL}}$$

$$\text{Seed correction, mg/L} = \frac{(\text{BOD}_5 \text{ of seed stock, mg/L})(\text{Seed stock, mg/L})}{\text{Total volume, mL}}$$

$\text{BOD}_5 \text{ seeded, mg/L} =$

$$\frac{(\text{Initial DO, mg/L} - \text{Final DO, mg/L} - \text{Seed correction, mg/L})(\text{Total volume, mL})}{\text{Sample volume, mL}}$$

BIOSOLIDS CONCENTRATION FACTOR

$$\text{CF} = \frac{\text{Sample volume, mL}}{\text{Percent influent biosolids}}$$

BIOSOLIDS PUMPING AND PRODUCTION FORMULAS

$\text{Estimated pumping rate} =$
$$\frac{(\text{Influent TSS, mg/L} - \text{Effluent TSS, mg/L})(\text{Flow, mgd})(8.34 \text{ lb/gal})}{(\text{Percent solids in sludge})(\text{Sludge, lb/gal})(1,440 \text{ min/day})}$$

$$\text{Biosolids, lb/mil gal} = \frac{(\text{Biosolids, gal})(8.34 \text{ lb/gal})}{(\text{Flow, mgd})(\text{Number of days})}$$

$$\text{Biosolids, lb/mil gal} = \frac{(\text{Biosolids, gal/day})(8.34 \text{ lb/gal})}{\text{Flow, mil gal}}$$

$$\text{Biosolids, wet tons/yr} = \frac{(\text{Biosolids, lb/mil gal})(\text{mgd})(365 \text{ days/yr})}{2,000 \text{ lb/ton}}$$

Estimated pumping rate =

$$\frac{(\text{Influent TSS, mg/L} - \text{Effluent TSS, mg/L})\,(\text{Flow, mgd})\,(8.34\,\text{lb/gal})}{(\text{Percent solids in sludge})\,(\text{Sludge, lb/gal})\,(1{,}440\,\text{min/day})}$$

BIOSOLIDS RETENTION TIME EQUATION

$$\text{BRT, days} = \frac{\text{Digester working volume, gal}}{\text{Influent flow, gpd}}$$

BIOSOLIDS VOLUME INDEX AND BIOSOLIDS DENSITY INDEX EQUATIONS

$$\text{BVI} = \frac{(\text{Settled biosolids, mL/L})\,(1{,}000\,\text{mg/g})}{\text{MLSS, mg/L}}$$

$$\text{BDI} = \frac{(\text{MLSS, mg/L})\,(100)}{(\text{Settled biosolids, mL/L})\,(1{,}000\,\text{mg/g})}$$

CENTRIFUGE THICKENING EQUATION

Hydraulic loading, gal/day = (Sludge flow, gpm)(1,440 min/day)

$$\text{Feed time, min} = \frac{(\text{Capacity, ft}^3)\,(\text{Solids, \%/100\%})\,(7.48\,\text{gal/ft}^3)\,(8.34\,\text{lb/gal})}{(\text{Flow, gpm})\,(\text{Solids concentration, \%/100\%})\,(8.34\,\text{lb/gal})}$$

$$\text{Simplified: Feed time, min} = \frac{(\text{Capacity, ft}^3)\,(\text{Solids, \%})\,(7.48\,\text{gal/ft}^3)}{(\text{Flow, gpm})\,(\text{Solids concentration, \%})}$$

CHEMICAL FEED SOLUTION SETTINGS

$$\text{Feed rate, mL/min} = \frac{(\text{gpd})\,(3{,}785\,\text{mL/gal})}{1{,}440\,\text{min/day}}$$

or

$$\text{Number of mL/min} = \frac{(\text{Number of gallons used})\,(3{,}785\,\text{mL/gal})}{1{,}440\,\text{min/day}}$$

CHEMICAL OXYGEN DEMAND LOADING FORMULA

COD, lb/day = (COD, mg/L)(Number of mgd)(8.34 lb/gal)

CHEMISTRY AND LABORATORY EQUATIONS

$$\text{Moles} = \frac{\text{Grams of chemical}}{\text{Gram formula weight}}$$

$$\text{Percent of element in compound} = \frac{(\text{Molecular Wt of the element})(100\%)}{\text{Molecular Wt of compound}}$$

$$\text{BOD}_5 \text{ unseeded, mg/L} = \frac{(\text{Initial DO, mg/L} - \text{Final DO, mg/L})(\text{Total volume, mL})}{\text{Sample volume, mL}}$$

$$\text{Seed correction, mg/L} = \frac{(\text{BOD}_5 \text{ of seed stock, mg/L})(\text{Seed stock, mg/L})}{\text{Total volume, mL}}$$

$\text{BOD}_5 \text{ seeded, mg/L} =$

$$\frac{(\text{Initial DO, mg/L} - \text{Final DO, mg/L} - \text{Seed correction, mg/L})(\text{Total volume, mL})}{\text{Sample volume, mL}}$$

$$\text{Percent of an element in a compound} = \frac{(\text{Molecular Wt of the element})(100\%)}{\text{Molecular weight of the compound}}$$

$$\text{Molarity} = \frac{\text{Moles solute}}{\text{Liters solution}}$$

$$\text{Normality (N)} = \frac{\text{Number of gram} - \text{equivalents of solute}}{\text{Number of liters of solution}}$$

$$\text{Dosage, mg/L} = \frac{(\text{Stock, mL})(1,000 \text{ mg/gram})(\text{Concentration in grams/liter})}{\text{Sample size, mL}}$$

$$\text{Percent VS} = \frac{(\text{Solids lost, g})(100\%)}{\text{Weight of total solids, g}}$$

$$\text{Alum, mg/L} = \frac{(\text{Stock, mL})(1,000 \text{ mg/gram})(\text{Concentration in grams/Liter})}{\text{Sample size, mL}}$$

Solids (ash), g = Sample and dish dried, g − Burnt sample, g

CIRCUMFERENCE FORMULAS

Circumference $= \pi(\text{Diameter})$

Circumference $= 2\pi(\text{radius})$ or $2\pi r$

COMMON CONVERSION FACTORS

Gallons to pounds: Number of pounds (lb) $=$ (Number of gal)(8.34 lb/gal)

Gallons to cubic feet: Number of $\text{ft}^3 = \dfrac{(\text{Number of gal})(1\ \text{ft}^3)}{7.48\ \text{gal}}$

Acre-feet to cubic feet: Number of $\text{ft}^3 =$ (Number of acre-ft)(43,560 ft^3/acre-ft)

mgd to ft^3/s: Number ft^3/s $=$ (Number of mgd)$\dfrac{(1,000,000\ \text{gal})}{(1\ \text{mil gal})}\ \dfrac{(1\ \text{ft}^3)}{(7.48\ \text{gal})}\ \dfrac{(1\ \text{day})}{(1,440\ \text{min})}\ \dfrac{(1\ \text{min})}{(60\ \text{sec})}$

ft^3/s to mgd: Number of mgd $= \dfrac{(\text{Number of ft}^3)}{\text{sec}}\ \dfrac{(60\ \text{sec})}{\text{min}}\ \dfrac{(1,440\ \text{min})}{\text{day}}\ \dfrac{(7.48\ \text{gal})}{\text{ft}^3}\ \dfrac{(1\ \text{mil gal})}{1,000,000\ \text{gal}}$

ft^3/s to gpd $=$ Number of gpd $=$ (Number of ft^3/s)(86,400 s/day)(7.48 gal/ft^3)

gpm to ft^3/s: Number of ft^3/s $= \dfrac{\text{Number of gpm}}{(60\ \text{s/min})(7.48\ \text{gal/ft}^3)}$

ppm to Percent: Percent solution $=$ (Known ppm) $\dfrac{(1\%)}{10,000\ \text{ppm}}$

Gallons to liters: Number of liters $=$ (Number of gal)(3.785 L/1 gal)

COMPOSTING EQUATIONS

Mixture's % moisture $= \dfrac{[(\text{Sludge, lb})(\%\ \text{moisture}) + (\text{Compost, lb})(\%\ \text{moisture})]100\%}{\text{Sludge, lb} + \text{Compost, lb}}$

Percent solids BC =

$$\frac{(\text{Sludge, yd}^3)(\text{lb/yd}^3)(\%\text{ solids, sludge}) + (\text{Sludge, yd}^3)(\text{MR})(\text{lb/yd}^3)(\%\text{ solids, chips})(100\%)}{(\text{Sludge, yd}^3)(\text{lb/yd}^3) + (\text{Sludge, yd}^3)(\text{Mix ratio})(\text{lb/yd}^3)}$$

$$\text{Compost cycle, days} = \frac{(\text{Site capacity, yd}^3)(\text{Density of compost, lb/yd}^3)}{x\text{ Wet compost, lb/day}}$$

Cycle time, days =

$$\frac{(\text{Capacity, yd}^3)(\text{Bulk density of compost lb/yd}^3)}{\dfrac{x\text{ Dry solids, lb/day}}{\text{Percent solids}} + \dfrac{(x\text{ Dry solids, lb/day})}{\text{Percent solids}}(\text{MR})\dfrac{(\text{Bulk density of wood chips, lb/yd}^3)}{(\text{Bulk density of wet sludge lb/yd}^3)}}$$

Percent moisture in mixture =

$$\frac{[(\text{DB, lb/day})(\text{Percent moisture DB}) + (\text{Compost lb/day})(\text{Percent moisture compost})]100\%}{\text{DB, lb/day} + \text{Compost, lb/day}}$$

DENSITY EQUATIONS

Density = Mass/Volume

Number of g/cm^3 = (Number of lb/gal)(454 g/1 lb)(1 gal/3,785 cm^3)

DETENTION TIME EQUATION

$$\text{Detention time, hr} = \frac{\text{Volume, gal}}{\text{Flow rate, gal/ hour}}$$

$$\text{Detention time, hr} = \frac{(\text{Volume, gal})(24\text{ hr/day})}{\text{Flow, gpd}}$$

DEWATERING FORMULAS

Total nonfilterable residue, mg/L = Total residue, mg/L − Total filterable residue, mg/L

$$\text{Filter yield, lb/hr/ft}^2 = \frac{(\text{Wet cake flow, lb/hr})(\text{Percent solids}/100\%)}{\text{Area, ft}^2}$$

$$\text{Vacuum filter loading, lb/day/ft}^2 = \frac{(\text{Biosolids, gpd})(8.34 \text{ lb/gal})(\text{Percent solids})}{\text{Vacuum filter area, ft}^2}$$

DIGESTER GAS PRODUCTION FORMULA

$$\text{Gas produced, ft}^3/\text{lb VS destroyed} = \frac{\text{Gas production, ft}^3/\text{day}}{\text{VS destroyed, lb/day}}$$

$$\text{Gas produced, m}^3/\text{lb VS destroyed} = \frac{(\text{Gas production, ft}^3/\text{day})}{(\text{VS destroyed, lb/day})(35.3 \text{ m}^3/\text{ft}^3)}$$

DIGESTER LOADING RATE EQUATION

$$\text{Digester loading rate, lb VSA/d/ft}^3 = \frac{\text{lb VSA}}{\text{Volume of digester, ft}^3}$$

Where VSA is Volatile Solids Added

Digester loading, lb VS/day/ft^3 =

$$\frac{(\text{Flow, gpd})(8.34 \text{ lb/gal})(\text{sp gr})(\text{Percent sludge})(\text{Percent volatile solids})}{(0.785)(\text{Diameter})^2(\text{Sludge level})}$$

Digester loading, lb VS/day/1,000 ft^3 =

$$\frac{(\text{Flow, gpd})(8.34 \text{ lb/gal})(\text{sp gr})(\text{Percent sludge})(\text{Percent volatile solids})}{(0.785)(\text{Diameter})^2(\text{Sludge level})}$$

DIGESTER VOLATILE SOLIDS RATIO FORMULA

$$\text{Digester VS ratio} = \frac{\text{VS added lb/day}}{\text{lb VS in digester}}$$

$$\text{Digester VS ratio} = \frac{\text{VS added lb/day}}{(\text{lb VS in digester})(\text{TS }\%/100\%)(\text{VS }\%/100\%)}$$

DILUTION TRIANGLE

Concentration (Conc.)$_1$

Conc. Desired

Conc.$_2$

Conc.$_2$ − Conc. Desired = Number of parts

Conc.$_1$ − Conc. Desired = $\dfrac{\text{Number of parts}}{\text{Total number of parts}}$

$$\frac{(\text{Number of gallons})\,(\text{Number of parts of Conc.}_1)}{\text{Total number of parts}} = \text{Number of gallons of Conc.}_1 \text{ required}$$

$$\frac{(\text{Number of gallons})\,(\text{Number of parts of Conc.}_2)}{\text{Total number of parts}} = \text{Number of gallons of Conc.}_2 \text{ required}$$

DISSOLVED AIR FLOTATION: AIR RATE FLOW EQUATIONS

Air, lb/day = (Air flow, ft^3/min)(1,440 min/day)(0.0807 lb/ft^3, Air)

Air, lb/day = (Air flow, ft^3/min)(60 min/hr)(0.0807 lb/ft^3, Air)

DISSOLVED AIR FLOTATION: AIR-TO-SOLIDS RATIO EQUATION

$$\text{Air-to-solids ratio} = \frac{(\text{Air flow, ft}^3/\text{min})\,(\text{Air, lb/ft}^3)}{(\text{gpm})\,(\text{Percent solids}/100\%)\,(8.34\ \text{lb/gal})}$$

DISSOLVED AIR FLOTATION: THICKENER SOLIDS LOADING EQUATIONS

$$\text{Solids loading, lb/d/ft}^2 = \frac{(\text{WAS, mg/L})\,(\text{mgd})\,(\text{Sludge, lb/gal})}{\text{DAF area, ft}^2}$$

$$\text{Solids loading, lb/hr/ft}^2 = \frac{(\text{WAS, mg/L})\,(\text{mgd})\,(\text{Sludge, lb/gal})}{(\text{DAF area, ft}^2)\,(24\ \text{hr/day})}$$

DOSAGE FORMULAS

Chlorine Dose = Chlorine demand + Chlorine residual

Chemical feed, lb/day = (Flow, mgd)(Dosage, mg/L)(8.34 lb/gal)

or rearranging to solve for dosage:

$$\text{Dosage, mg/L} = \frac{\text{lb/day}}{\text{(mgd)}\,(8.34\text{ lb/gal})}$$

$$\text{lb/day} = \frac{\text{(mgd)}\,\text{(Dosage, mg/L)}\,(8.34\text{ lb/gal})}{\text{(Percent purity/100\%)}}$$

Above formula used when the purity of a substance or solution is less than 100%.

$$(\text{mgd})(x,\text{ mg/L})(8.34\text{ lb/gal}) = (\text{mgd})(\text{Dosage, mg/L})(8.34\text{ lb/gal})$$

$$\text{Chlorine, lb/day} = \frac{\text{(Dosage, mg/L)}\,\text{(mgd)}\,(8.34\text{ lb/gal})}{\text{Percent available chlorine/100\%}}$$

$$\text{Chemical dosage, mg/L} = \frac{\text{(mL/min)}\,(1{,}440\text{ min/day})\,\text{(Chemical, lb/gal)}}{(3{,}785\text{ mL/gal})\,\text{(mgd)}\,(8.34\text{ lb/gal})}$$

$$\text{Chemical dosage, mg/L} = \frac{\text{(mL/min)}\,(1{,}440\text{ min/day})\,\text{(Chemical, lb/gal)}}{(3{,}785\text{ mL/gal})\,\text{(mgd)}\,(8.34\text{ lb/gal})\,\text{(Percent Polymer)}}$$

DRY CHEMICAL FEED SETTINGS

$$\text{Chemical, lb/day} = \frac{\text{(Number of g/min)}\,(1{,}440\text{ min/day})}{454\text{ g/lb}}$$

EXTRAPOLATION: USED FOR PIPES NOT FLOWING FULL

$$\text{Division Factor} = \frac{\text{(High d/D} - \text{Low d/D)}}{\text{Ratio d/D}}$$

FOOD/MICROORGANISM RATIO FORMULA

$$F/M = \frac{(BOD_5, mg/L)(Flow, mgd)(8.34 \, lb/gal)}{(mg/L \, MLVSS)(Volume \, in \, tank, \, mil \, gal)(8.34 \, lb/gal)}$$

or as follows since 8.34 lb/gal in the numerator and denominator cancel each other out.

$$F/M = \frac{(BOD_5, mg/L)(Flow, mgd)}{(mg/L \, MLVSS)(Volume \, of \, tank, \, mil \, gal)}$$

FLOW RATE EQUATIONS

Flow = Volume/Time

Q (Flow) = (Area)(Velocity)

Example: Q, flow in ft³/s = (Area, ft²)(Velocity in feet per sec)

Flow in a pipe that changes size: (Area 1, sq ft)(Velocity 1, ft/s) = (Area 2, sq ft)(Velocity 2, ft/s)

$$\text{Flow (Q), ft}^3/\text{sec} = \frac{(w_1 + w_2)}{2}(\text{Depth, ft})(\text{Velocity, ft/s})$$

GRAVITY THICKENER SOLIDS LOADING FORMULA

$$\text{Solids loading, lb/d/ft}^2 = \frac{(\text{Flow, gpm})(1,440 \, \text{min}/\text{day})(\text{Percent solids})}{(\text{Gravity thickener area})(100\%)}$$

$$\text{Solids loading, lb/d/ft}^2 = \frac{(\text{Flow, gpd})(\text{Percent solids})}{(\text{Gravity thickener area})(100\%)}$$

GRIT REMOVAL FORMULA

$$\text{Grit removal, ft}^3/\text{mil gal} = \frac{\text{Number of gallons removed}}{(7.48 \, \text{gal/ft}^3)(\text{mil gal treated})}$$

HYDRAULIC DIGESTION TIME EQUATION

$$\text{Digestion time, days} = \frac{\text{Number of gallons}}{\text{Influent sludge flow, gal/day}}$$

$$\text{Digestion time, days} = \frac{(0.785)(\text{Diameter})^2(\text{Depth, ft})(7.48\ \text{gal/ft}^3)}{\text{Influent sludge flow, gal/day}}$$

HYDRAULIC LOADING RATE EQUATION

$$\text{Hydraulic loading rate} = \frac{\text{Total flow, gpd}}{\text{Surface area, ft}^2}$$

$$\text{Hydraulic loading rate, in/day} = \frac{\text{Flow, gpd}}{(27{,}152\ \text{gal/acre-in.})(\text{Area, acres})}$$

KILOWATT FORMULAS

$$\text{kW} = (\text{Number of hp})(0.746\ \text{kW/hp})$$

$$\text{Kilowatt-hr/day} = (\text{hp})(0.746\ \text{kW/hp})(\text{Operating time, hr/day})$$

$$\text{kW} = (\text{Number of hp})(0.746\ \text{kW/hp})(\text{Startup energy})$$

LIME NEUTRALIZATION FORMULA

$$\text{Lime, lb} = (\text{Volatile acids, mg/L})(\text{mil gal})(8.34\ \text{lb/gal})$$

MASS BALANCE (PERCENT) EQUATION

$$\text{Percent mass balance} = \frac{(\text{Solids produced, lb/day} + \text{Solids removed, lb/day})(100\%)}{\text{Solids produced, lb/day}}$$

MEAN CELL RESIDENCE TIME EQUATIONS

$$\text{MCRT, days} = \frac{(\text{MLSS, mg/L})(\text{mil gal})(8.34 \text{ lb/day})}{\text{SS wasted, lb/day} + \text{SS lb/day}}$$

$$\text{MCRT, days} =$$

$$\frac{(\text{MLSS, mg/L})(\text{Aeration tank mil gal} + \text{Clarifier tank mil gal})(8.34 \text{ lb/gal})}{(\text{WAS, mg/L})(\text{Waste rate, mgd})(8.34 \text{ lb/gal}) + (\text{TSS, mg/L})(\text{Flow, mgd})(8.34 \text{ lb/gal})}$$

MIXTURE FORMULA

Percent mixture strength $=$

$$\frac{\text{Solution 1 gal}(\text{Available \%}/100\%) + \text{Solution 2 gal}(\text{Available \%}/100\%)(100\%)}{\text{gal of solution 1} + \text{gal of solution 2}}$$

NITROGEN LOADING RATE EQUATION

Nitrogen loading rate, lb/day $=$ (Total Nitrogen, mg/L)(mgd)(8.34 lb/gal)

NITROGEN (TOTAL) EQUATION

Total nitrogen (N) = Nitrate, mg/L + Nitrite, mg/L + TKN, mg/L

OPERATING TIME FORMULAS

$$\text{Operating time} = \frac{\text{Treated water}}{\text{Flow rate}}$$

$$\text{Operating time, min/hr} = \frac{(\text{Flow, mgd})(\text{Influent SS, mg/L} - \text{Effluent SS, mg/L})(100\%)}{(\text{Sludge pump, gpm})(\text{Percent solids})(24 \text{ hr/day})}$$

ORGANIC LOADING RATE EQUATIONS

$$\text{Organic loading rate, lb BOD}_5\text{/d/acre} = \frac{(\text{BOD}_5, \text{mg/L}) \, (\text{Flow, mgd}) \, (8.34 \, \text{lb/gal})}{\text{Surface area of pond, acre-ft}}$$

$$\text{Organic loading rate, lb BOD}_5\text{/d/1,000 ft}^3 = \frac{(\text{BOD}_5, \text{mg/L}) \, (\text{Flow, mgd}) \, (8.34 \, \text{lb/gal})}{\text{Volume of trickling filter, ft}^3/1,000 \, \text{ft}^3}$$

PARTICULATE AND SOLUBLE BIOLOGICAL OXYGEN DEMAND FORMULAS

Particulate BOD_5, mg/L = (SS, mg/L)(K value)

Soluble BOD_5 = Total BOD_5 − (K factor)(Total SS)

Total BOD_5 = (Particulate BOD_5)(K factor) + Soluble BOD_5

PERIMETER FORMULAS

Circumference = π(Diameter) = (3.14)(Diameter)

Rectangle = 2(Length) + 2(Width)

PLANT AVAILABLE NITROGEN FORMULAS

PAN, mg/L = [MR(TKN − NH_3)] + [0.50(NH_3)] + (NO_3 + NO_2)

$$\text{PAN, dry tons/acre} = \frac{\text{Plant nitrogen required, lb/acre}}{\text{PAN, lb/dry ton}}$$

PAN, lb/dry ton = [(Organic N, mg/kg)(MR) + (Ammonia N, mg/kg)(VR)](0.002 lb/dry ton)

POPULATION EQUIVALENT EQUATION

$$\text{Number of people} = \frac{(\text{BOD}_5, \text{mg/L})\,(\text{mgd})\,(8.34\ \text{lb/gal})}{\text{lb/day of BOD}_5/\text{person}}$$

POPULATION LOADING EQUATION

$$\text{Population loading, people/acre} = \frac{\text{Number of people served}}{\text{Area of pond(s), acres}}$$

PRESSURE FORMULAS

$$\text{psi} = \frac{\text{Depth, ft}}{2.31\ \text{ft/psi}}$$

$$\text{Height, ft} = (\text{psi})(2.31\ \text{ft/psi})$$

$$\text{psi} = (\text{Depth, ft})(0.433\ \text{psi/ft})$$

$$\text{Pressure} = \frac{\text{Force, lb}}{\text{Area, ft}^2}$$

$$\text{Pressure, lb/ft}^2 = (\text{Height or Depth, ft})(\text{Density, 62.4 lb/ft}^3)$$

$$\frac{\text{Pressure}_A}{w} + \frac{\text{Velocity}_A^{\,2}}{2g} = \frac{\text{Pressure}_B}{w} + \frac{\text{Velocity}_B^{\,2}}{2g}$$

PERCENT MIXTURE FORMULAS

Percent mixture strength =

$$\frac{(\text{Solution}_1\ \text{lb})(\text{Available \%}/100\%) + (\text{Solution}_2,\ \text{lb})(\text{Available \%}/100\%)(100\%)}{\text{Solution}_1,\ \text{lb} + \text{Solution}_2,\ \text{lb}}$$

Percent mixture strength =

$$\frac{[(\text{Solution}_1\ \text{gal})(\text{lb/gal})(\text{Avail \%}/100\%) + (\text{Solution}_2\ \text{gal})(\text{lb/gal})(\text{Avail \%}/100\%)]100\%}{(\text{Solution}_1,\ \text{gal})(\text{lb/gal}) + (\text{Solution}_2,\ \text{gal})(\text{lb/gal})}$$

PERCENT RECOVERY EQUATION

$$\text{Percent recovery} = \frac{\text{Cake TS, \%}(\text{Feed sludge TSS, \%} - \text{Return flow TSS, \%})(100\%)}{\text{Feed sludge TSS, \%}(\text{Cake TS, \%} - \text{Return flow TSS, \%})}$$

PERCENT REDUCTION EQUATIONS

$$\text{Percent VS reduction} = \frac{(\text{Influent} - \text{Effluent})(100\%)}{\text{Effluent} - (\text{Effluent})(\text{Influent})}$$

$$\text{Percent VM reduction} = \frac{(\text{Percent influent VM} - \text{Percent effluent VM})(100\%)}{[\text{Percent influent VM} - (\text{Percent influent VM})(\text{Percent effluent VM})]}$$

Percent moisture reduction =

$$\frac{(\text{Percent influent moisture} - \text{Percent moisture, after digestion})(100\%)}{[\text{Percent influent moisture} - (\text{Percent influent moisture})(\text{Percent moisture, after digestion})]}$$

PERCENT REMOVAL FORMULA

$$\text{Percent NTU removal} = \frac{(\text{Influent ntu} - \text{Effluent ntu})(100\%)}{\text{Influent ntu}} \text{ or } \frac{(\text{In} - \text{Out})(100\%)}{\text{In}}$$

$$\text{Percent BOD}_5 \text{ removal} = \frac{(\text{Influent BOD}_5 - \text{Effluent BOD}_5)(100\%)}{\text{Influent BOD}_5} \text{ or } \frac{(\text{In} - \text{Out})(100\%)}{\text{In}}$$

$$\text{Percent removal efficiency} = \frac{(\text{Solids removed, mg/L})(100\%)}{\text{Influent solids, mg/L}}$$

PERCENT STRENGTH OF SOLUTIONS AND SOLIDS

$$\text{Percent strength} = \frac{(\text{Number of lb of chemical})(100\%)}{\text{Number of lb, Water} + \text{lb chemical}}$$

$$\text{Percent strength} = \frac{(\text{Number of grams of chemical})(100\%)}{\text{Number of grams, water} + \text{grams chemical}}$$

$$\text{Percent total solids} = \frac{(\text{Dry sample in grams})(100\%)}{\text{Sludge sample in grams}}$$

PERCENT SETTLED SLUDGE AND SOLIDS FORMULAS

$$\text{Percent settled sludge} = \frac{(\text{Settled sludge, mL})(100\%)}{\text{Total sample vol., mL}}$$

or similarly:

$$\text{Percent settleable solids} = \frac{(\text{Settled sludge, mL})(100\%)}{\text{Sample size, mL}}$$

$$\text{Percent inorganic solids} = \frac{(\text{Dry sample in grams})(100\%)}{\text{Sludge sample in grams}}$$

$$(2° \text{ gpd})(2° \text{ sludge lb/gal})(x\% \ 2° \text{ sludge}) = (1° \text{ sludge, gpd})(1° \text{ sludge lb/gal})(\% \ 1° \text{ sludge})$$

PHOSPHATE LOADING RATE EQUATION

$$\text{Phosphorus (P) loading rate, lb/day} = (\text{P, mg/L})(\text{mgd})(8.34 \text{ lb/gal})$$

PIT VOLUME AND DAYS TO FILL FORMULAS

$$\text{Screenings, ft}^3/\text{mil gal} = \frac{\text{Number of ft}^3/\text{day}}{\text{Number of mgd}}$$

$$\text{Number of days to fill} = \frac{\text{Pit volume, ft}^3}{\text{Screenings removed, ft}^3/\text{day}}$$

PUMPING AND PUMPING COST FORMULAS

$$\text{mhp} = \frac{(\text{whp})}{(\text{Motor efficiency})(\text{Pump efficiency})}$$

Where mhp = motor horsepower and whp = water horsepower

$$\text{mhp} = \frac{(\text{Flow, gpm})(\text{TH, ft})}{(3,960)(\text{Motor efficiency})(\text{Pump efficiency})}$$

Where TH = Total head

$$bhp = \frac{(Flow, gpm)(Differential\ pressure, psi)}{(1,714)(Pump\ efficiency)}$$

Brake hp = (hp)(Motor efficiency)

Where hp = horsepower

Water hp = (mhp)(Motor efficiency)(Pump efficiency)

Water hp = (bhp)(Pump efficiency)

Motor hp = bhp/Motor efficiency

Brake hp = whp/Pump efficiency

Cost, $/day = (Motor hp)(24 hr/day)(0.746 kW/hp)(Cost/kW-hr)

PUMPING RATE EQUATIONS

Pumping rate = Flow, gal/Time, min

$$Pump's\ discharge\ rate, gpm = \frac{Discharge, gal}{Time, min}$$

Discharge rate, gpm = Influent flow, gpm + Level drop, gpm

Number of gal per stroke = (0.785)(Bore diameter, ft)^2(Stroke, ft)(7.48 gal/ft^3)

RATIOS

$$\text{Ratio} = \frac{\text{Re circulated flow}}{\text{Plant influent flow}}$$

$$\frac{\text{Speed setting}_1, \text{Percent}}{\text{Polymer dosage}_1, \text{mL}} = \frac{\text{Speed setting}_2, \text{Percent}}{\text{Polymer dosage}_2, \text{mL}}$$

$$\frac{\text{Chlorine dosage}_1, \text{mg/L}}{\text{Flow}_1, \text{mgd}} = \frac{\text{Chlorine dosage}_2, \text{mg/L}}{\text{Flow}_2, \text{mgd}}$$

$$\frac{\text{Digester solids}_1, \text{lb/day}}{\text{Flow}_1, \text{gpm}} = \frac{\text{Digester solids}_2, \text{lb/day}}{\text{Flow}_2, \text{gpm}}$$

$$\frac{\text{Alum dosage}_1, \text{mL}}{\text{Speed setting}_1, \%} = \frac{\text{Alum dosage}_2, \text{mL}}{\text{Speed setting}_2, \%}$$

SCREENINGS FORMULA

$$\text{Screenings, ft}^3/\text{mil gal} = \frac{\text{Number of ft}^3/\text{day}}{\text{Number of mgd}}$$

SEED SLUDGE EQUATION

$$\text{Seed sludge, gal} = \frac{(\text{Capacity of digester})(\text{Percent seed sludge required})}{100\%}$$

SLUDGE AGE (GOULD) EQUATION

$$\text{Sludge age, days} = \frac{\text{Solids under aeration, lb}}{\text{Solids added, lb/day}}$$

$$\text{Sludge age, days} = \frac{(\text{MLSS, mg/L})(\text{Volume of aeration tank})(8.34 \text{ lb/gal})}{(\text{SS, mg/L})(\text{Flow, mgd})(8.34 \text{ lb/gal})}$$

SLUDGE PUMPING FORMULA

Sludge, lb/day = (Pumping, min/day)(24 hr/day)(Pump rate, gpm)(8.34 lb/gal)(sp gr of sludge)

(Primary sludge, gal)(Primary sludge, lb/gal)(Percent PSS) =

(x Thickened sludge, gal)(Thickened sludge, lb/gal)(Percent TSS)

SLUDGE REMOVED EQUATION

SS removed, lb/day = (SS removed, mg/L)(Number of mgd)(8.34 lb/gal)

SLUDGE VOLUME INDEX AND SLUDGE VOLUME DENSITY EQUATIONS

$$SVI = \frac{(SS, mL)}{MLSS, g/L}$$

$$SDI = \frac{(MLSS, g)(100\%)}{SS, mL}$$

SODIUM ABSORPTION RATIO FORMULA

$$Sodium\ absorption\ ratio = \frac{Na^+}{[(0.5)(Ca^{2+} + Mg^{2+})]^{1/2}}$$

SOLIDS BALANCE (DIGESTER) EQUATIONS

Total solids, lb/day = (Raw sludge, lb/day)(Percent solids)

Fixed solids, lb/day = Total solids, lb/day − VS, lb/day

Water in sludge, lb/day = Sludge, lb/day − Total solids, lb/day

$$\text{Percent VSR} = \frac{(\text{In} - \text{Out})(100\%)}{\text{In} - (\text{In})(\text{Out})}$$

Gas produced, lb/day = (Effluent VS, lb/day)(Percent VSR)

VS in digested sludge, lb/day = Influent VS, lb/day − Destroyed VS, lb/day

$$\text{Total digested solids, lb/day} = \frac{\text{VS digested, lb/day}}{\text{Percent digested VS}}$$

Fixed solids, lb/day = Total digested solids, lb/day − VS digested, lb/day

$$\text{Digested sludge, lb/day} = \frac{\text{Total digested solids, lb/day}}{\text{Digested sludge percent solids}}$$

Water in digested sludge, lb/day = Sludge, lb/day − Total solids, lb/day

SOLIDS LOADING RATE EQUATIONS

$$\text{Solids loading rate} = \frac{(\text{MLSS, mg/L})(\text{mgd})(8.34\ \text{lb/gal})}{\text{Area, ft}^2}$$

$$\text{Solids loading rate, lb/d/ft}^2 = \frac{(\text{Percent solids})(\text{Biosolids added, gpd})(8.34\ \text{lb/gal})}{(\text{Surface area, ft}^2)}$$

SOLIDS PRODUCED FORMULA

Solids produced, lb/day = (BOD$_5$ removed, lb/day)(0.85 lb solids/lb BOD$_5$)

SOLIDS PUMPING EQUATIONS

$$\text{Solids, lb/day} = \frac{(\text{Sludge, lb/day}) (\text{Percent solids})}{100\%}$$

$$\text{Solids, lb/day} = (\text{Pumping, min/day})(24 \text{ hr/day})(\text{Pump rate, gpm})(8.34 \text{ lb/gal})(\text{sp gr of sludge})$$

$$\text{Solids, lb/day} = (\text{Time, min/cycle})(\text{cycles/day})(\text{Pump rate, gpm})(8.34 \text{ lb/gal})(\text{Percent solids})$$

SOLIDS RETENTION TIME EQUATION

$$\text{Target SRT} =$$

$$\frac{(\text{MLSS mg/L}) (\text{Clarifier, Aerator Volume, mil gal}) (8.34 \text{ lb/gal})}{(\text{RAS SS mg/L}) (x \text{ mgd}) (8.34 \text{ lb/gal}) + (\text{Effluent SS, mg/L}) (\text{Flow, mgd}) (8.34 \text{ lb/gal})}$$

SOLIDS UNDER AERATION

$$\text{Number of lb, solids} = (\text{SS, mg/L})(\text{Number of mil gal})(8.34 \text{ lb/gal})$$

SOLUTION FORMULAS

$$(\text{Concentration}_1)(\text{Volume}_1) + (\text{Concentration}_2)(\text{Volume}_2) = (\text{Concentration}_3)(\text{Volume}_3)$$

or abbreviating the above equation: $C_1V_1 + C_2V_2 = C_3V_3$

$$\text{Percent HTH solution} = \frac{(\text{lb HTH}) (100\%)}{(\text{Number of gal}) (8.34 \text{ lb/gal})}$$

$(\text{Solution}_1 \text{ percent})(x \text{ gal, Solution}_1)(\text{Solution}_1, \text{lb/gal}) = (\text{Solution}_2 \text{ percent})(\text{Solution}_2, \text{gal})$
$(\text{Solution}_2, \text{lb/gal})$

SPECIFIC GRAVITY EQUATIONS

Specific gravity (sp gr) = Density of substance/Density of water

$$\text{Sp gr} = \frac{\text{Solute, lb/gal}}{8.34 \text{ lb/gal}}$$

$$\text{Sp gr} = \frac{\text{Number of lb/ft}^3}{62.4 \text{ lb/ft}^3}$$

STATISTIC FORMULAS

$$\text{Average} = \frac{\text{Sum of all measurements}}{\text{Number of measurements}}$$

or

$$\text{Arithmetic mean} = \frac{\text{Sum of all measurements}}{\text{Number of measurements}}$$

Median is the middle value.

Range = Largest value − Smallest value

Mode is the valve that occurs most frequently.

Note: There also can be two or more modes within a set of numbers.

Geometric mean = $[(x_1)(x_2)(x_3)(x_4).....(x_n)]^{1/n}$

Standard deviation = $[\Sigma f(x - X^-)^2/n - 1]^{1/2}$

SURFACE LOADING FORMULA

$$\text{Surface loading rate} = \frac{\text{gallons per day (gpd)}}{\text{Number of ft}^2}$$

SUSPENDED SOLIDS LOADING EQUATION

Suspended solids, lb/day = (SS, mg/L)(Number of mgd)(8.34 lb/gal)

TEMPERATURE FORMULAS

$$°C = 5°C / 9°F \ (°F - 32)$$

$$°F = 9°F / 5°C \ (°C) + 32°F$$

TOTAL FORCE AND HYDRAULIC PRESS EQUATIONS

Total force, pounds = (Area, in.^{2})(psig) or Total force = (Pressure)(Area)

$$\text{Pressure} = \frac{\text{Total force, lb}}{\text{Area, ft}^2}$$

TOTAL HEAD FORMULAS

Total head, ft = Total static head, ft + Head losses, ft

$$\text{TDH} = \frac{(\text{Differential pressure}) \, (2.31 \text{ ft/psi})}{\text{Specific gravity}}$$

VACUUM FILTER OPERATING TIME EQUATION

$$\text{Filter yield, lb/hr/ft}^2 = \frac{\dfrac{(\text{Solids, lb/day}) \, (\text{Percent recovery})}{(\text{Filter operation, lb/day}) \, (100\%)}}{\text{Filter area ft}^2}$$

VELOCITY EQUATIONS

$$\text{Velocity, ft/s} = \frac{\text{Flow, gpm}}{(\text{Width, ft})(\text{Depth, ft})(60 \text{ sec}/\text{min})(7.48 \text{ gal/ft}^3)}$$

or

$$\text{Velocity, ft/s} = \frac{\text{Flow, gpm}}{(\text{Area, ft}^2)(60 \text{ sec}/\text{min})(7.48 \text{ gal/ft}^3)}$$

VOLATILE ACIDS-TO-ALKALINITY RATIO FORMULAS

Ratio = Volatile acids/Alkalinity

Volatile acids = (Alkalinity)(Ratio)

$$\text{Digester VS ratio} = \frac{\text{VS added lb/day}}{\text{lb VS in digester}}$$

VOLATILE SOLIDS DESTROYED EQUATION

$$\text{VS destroyed, lb/day/ft}^3 = \frac{(\text{Flow, gpd})(\text{Sludge, lb/gal})(\text{SSC, \%})(\text{VSC, \%})(\text{VSR, \%})}{\text{Digester capacity, 1,000 ft}^3}$$

VOLATILE SOLIDS: LB/DAY (PERCENT)

VS, lb/day = (Number of lb/day, sent to digester)(Percent VS/100%)

$$\text{VS, lb/day} = (\text{Number of gpd to digester})\frac{(\text{Percent solids})}{100\%}\frac{(\text{Percent VS})}{100\%}(8.34 \text{ lb/gal})$$

VOLATILE SOLIDS PUMPING FORMULA

VS, lb/day =

(Time, min/cycle)(cycles/day)(Pump rate, gpm)(8.34 lb/gal)(Percent, solids)(Percent VM)

VOLUME EQUATIONS

Volume of a basin in ft^3 or m^3 = (Length)(Width)(Depth)

Volume of a basin in gallons = (Length)(Width)(Depth)(7.48 gal/ft^3)

Volume of a cone in ft^3 or m^3 = $1/3\pi r^2$(Height or Depth)

Volume of a circular tank in ft^3 or m^3 = πr^2(Height) or Volume of a pipe = πr^2(Length) or

Volume of a cylindrical tank in ft^3 or m^3 = (0.785)(Diameter)^2(Height)

Volume of a trough in gallons = $\dfrac{(b_1 + b_2)}{2}$(Depth of water)(Length)(7.48 gal/ft^3)

Volume, gal = $\dfrac{(Length_1 + Length_2)}{2} \dfrac{(Width_1 + Width_2)}{2}$(Depth, ft)(7.48 gal/ft^3)

Volume of sphere, ft^3 = $\dfrac{4\pi r^3}{3}$

Digester capacity, ft^3 = π(radius)^2(Height, ft)

Equation below is for a partially filled pipe division factor for determining flow:

Division factor = $\dfrac{(High\ d/D - Low\ d/D)}{Ratio\ d/D}$

WASTE ACTIVATED SLUDGE LOADING RATE EQUATION

WAS, lb/day = (WAS, mg/L)(Number of mgd)(8.34 lb/gal)

WASTE ACTIVATED SLUDGE PUMPING RATE FORMULA

$$\text{Number of mgd} = \frac{\text{WAS, lb/day}}{(\text{Number of mg/L WAS})(8.34 \text{ lb/gal})}$$

WASTE RATE EQUATION

Waste rate, mgd = Waste, lb

(WVS concentration, mg/L)(8.34 lb/gal)

The following two equations are used to find the waste rate (above equation).

$$\text{Desired MLVSS, lb} = \frac{(\text{Primary effluent COD, mg/L})(\text{mgd})(8.34 \text{ lb/gal})}{\text{Desired COD lb/MLVSS lb}}$$

Existing MLVSS, lb = (MLVSS, mg/L)(Aeration tank, mil gal)(8.34 lb/gal)

$$\text{Waste rate, gpd} = \frac{(\text{Solids produced, lb/day})(1,000,000/\text{mil})}{(\text{Waste TSS, mg/L})(8.34 \text{ lb/gal})}$$

Waste rate, lb/day =

$$\frac{\text{MLSS, mg/L}\,[\text{AT, mil gal} + \text{CT, mil gal}]\,(8.34 \text{ lb/gal})}{\text{Desired MCRT}} - (\text{TSS, mg/L})(\text{Flow, mgd})(8.34 \text{ lb/gal})$$

WEIR LENGTH FORMULA

Weir length, ft $= \pi(\text{Diameter, ft})$

Weir length, ft $= 2\pi(\text{radius, ft})$

WEIR AND SURFACE OVERFLOW RATE EQUATIONS

Weir overflow rate $= \dfrac{\text{Flow, gpd}}{\text{Weir Length, ft}}$

Surface overflow rate $= \dfrac{\text{Flow, gpd}}{\text{Area, ft}^2}$

APPENDIX C

CHEMISTRY TABLES

TABLE C-1 International Atomic Weights (Based on Carbon-12)							
Element	Symbol	Atomic Number	Atomic Weight	Element	Symbol	Atomic Number	Atomic Weight
Actinium	Ac	89	227.0278	Magnesium	Mg	12	24.305
Aluminum	Al	13	26.98154	Manganese	Mn	25	54.9380
Americium	Am	95	(243)*	Mendelevium	Mv	101	(258)
Antimony	Sb	51	121.75	Meitnerium	Mt	109	(266.0)
Argon	Ar	18	39.948	Mercury	Hg	80	200.59
Arsenic	As	33	74.9216	Molybdenum	Mo	42	95.94
Astatine	At	85	(210)	Neodymium	Nd	60	144.24
Barium	Ba	56	137.33	Neon	Ne	10	20.179
Berkelium	Bk	97	(247)	Neptunium	Np	93	237.0482
Beryllium	Be	4	9.01218	Nickel	Ni	28	58.69
Bismuth	Bi	83	208.9804	Niobium	Nb	41	92.9064
Bohrium	Bh	107	(262.0)	Nitrogen	N	7	14.0067
Boron	B	5	10.81	Nobelium	No	102	(259)
Bromine	Br	35	79.904	Osmium	Os	76	190.2
Cadmium	Cd	48	112.41	Oxygen	O	8	15.9994
Calcium	Ca	20	40.08	Palladium	Pd	46	106.42
Californium	Cf	98	(251)	Phosphorus	P	15	30.97376
Carbon	C	6	12.011	Platinum	Pt	78	195.08
Cerium	Ce	58	140.12	Plutonium	Pu	94	(244)
Cesium	Cs	55	132.9054	Polonium	Po	84	(209)

*The numbers in parentheses indicate mass number of most stable known isotope.

Chlorine	Cl	17	35.453	Potassium	K	19	39.0983
Chromium	Cr	24	51.996	Praseodymium	Pr	59	140.9077
Cobalt	Co	27	58.9332	Promethium	Pm	61	(145)
Copper	Cu	29	63.546	Protactinium	Pa	91	2310359
Curium	Cm	96	(247)*	Radium	Ra	88	226.0254
Dubnium	Db	105	(262.0)	Radon	Rn	86	(222)
Dysprosium	Dy	66	162.50	Rhenium	Re	75	186.207
Einsteinium	Es	99	(252)	Rhodium	Rh	45	102.9055
Erbium	Er	68	167.26	Rubidium	Rb	37	85.4678
Europium	Eu	63	151.96	Ruthenium	Ru	44	101.07
Fermium	Fm	100	(257)	Rutherfordium	Rf	104	(261.0)
Fluorine	F	9	18.998403	Samarium	Sm	62	150.36
Francium	Fr	87	(223)	Scandium	Sc	21	44.9559
Gadolinium	Gd	64	157.25	Seaborgium	Sg	106	(263.0)
Gallium	Ga	31	69.72	Selenium	Se	34	78.96
Germanium	Ge	32	72.59	Silicon	Si	14	28.0855
Gold	Au	79	196.9665	Silver	Ag	47	107.8682
Hafnium	Hf	72	178.49	Sodium	Na	11	22.98977
Hassium	Hs	108	(265.0)	Strontium	Sr	38	87.62
Helium	He	2	4.00260	Sulfur	S	16	32.06
Holmium	Ho	67	164.9304	Tantalum	Ta	73	180.9479
Hydrogen	H	1	1.00794	Technetium	Tc	43	(98)
Indium	In	49	114.82	Tellurium	Te	52	127.60
Iodine	I	53	126.9045	Terbium	Tb	65	158.9254
Iridium	Ir	77	192.22	Thallium	Tl	81	204.383
Iron	Fe	26	55.847	Thorium	Th	90	232.0381
Krypton	Kr	36	83.80	Thulium	Tm	69	168.9342
Lanthanum	La	57	138.9055	Tin	Sn	50	118.69
Lawrencium	Lw	103	(260)	Titanium	Ti	22	47.88
Lead	Pb	82	207.2	Tungsten	W	74	183.85
Lithium	Li	3	6.941	Ununnilium	Uun	110	(269.0)
Lutetium	Lu	71	174.967	Unununium	Uuu	111	(272.0)
Unumbium	Uub	112	(277.0)	Ytterbium	Yb	70	173.04
Uranium	U	92	238.0289	Yttrium	Y	39	88.9059
Vanadium	V	23	50.9415	Zinc	Zn	30	65.38
Xenon	Xe	54	131.29	Zirconium	Zr	40	91

Source: US Government Printing Office
*The numbers in parentheses indicate mass number of most stable known isotope.

TABLE C-2			
Common Chemicals Used and Encountered in the Water Industry			
Chemical	**Chemical Formula**	**Use**	**Miscellaneous**
Alum	$Al_2(SO_4)_3 \cdot 14(H_2O)$	Coagulant	
Ammonia	NH_3	Check for chlorine leaks	Make chloramine
Bicarbonate	HCO_3^-		Carbonate hardness
Calcium carbonate	$CaCO_3$	Primary hardness chemical	
Calcium hypochlorite	$Ca(OCl)_2$	Disinfectant	
Carbon dioxide	CO_2	Re-carbonation	
Caustic soda	$NaOH$	Adjust pH	
Chlorine	Cl_2	Disinfectant	
Chlorine dioxide	ClO_2	Disinfectant	
Copper sulfate	$CuSO_4$	Algae control	
Dichloramine	$NHCl_2$	Disinfectant	Formed when ammonia is and to water containing Cl
Ferric chloride	$FeCl_3$	Coagulant	
Ferric sulfate	$Fe_2(SO_4)_3$	Coagulant	
Ferrous sulfate	$Fe_2(SO_4)_3 \cdot 7(H_2O)$	Coagulant	
Fluorosilicic acid*	H_2SiF_6	Fluoridation	
Hydrated lime	$Ca(OH)_2$	Increase pH and alkalinity	
Hydrochloric acid	HCl	Laboratory, cleaning	
Hydroxide ion	OH^-		Naturally found in water
Hypochlorite ion	OCl^-	Disinfectant	
Magnesium hydroxide	$Mg(OH)_2$		Formed in the lime-soda softening process
Monochloramine	NH_2Cl	Disinfectant	Formed when ammonia is and to water containing Cl
Nitrate	NO_3^-		Fertilizer, sewage, natural deposits that are eroded
Quicklime	CaO	Water stabilization, increase pH and alkalinity	Water softening
Ozone	O_3	Disinfectant	
Potassium permanganate	$KMnO_4$	Control tastes-and-odors substances	Oxidize Fe and Mn
Sulfuric acid	H_2SO_4	Decrease pH and alkalinity	Water stabilization
Soda ash	Na_2CO_3	Adjust pH	
Sodium aluminate	$Na_2Al_2O_4$	Coagulant	
Sodium bicarbonate	$NaHCO_3$	Increase pH and alkalinity	Water stabilization
* Formerly known as hydrofluosilicic acid or "silly acid."			

Chemical	Chemical Formula	Use	Miscellaneous
Sodium fluoride	NaF	Fluoridation	
Sodium hexametaphosphate	$(NaPO_3)_n \cdot Na_2O$*	Sequestering agent	
Sodium hydroxide	$NaOH$	Adjust pH	
Sodium silicate	$Na_2O \cdot (SiO_2)_x$	Coagulant	
Sodium fluorosilicate[†]	Na_2SiF_6	Fluoridation	
Trichloramine	NCl_3		Formed when ammonia is and to water containing Cl
Zinc orthophosphate	$Zn_3(PO_4)_2$	Forms protective coating	
* Typically n = 14 † Formerly known as sodium silicofluoride.			

TABLE C-3
Common Formula Weights*

Compound	Weight in Grams	Compound	Weight in Grams
$AgCl$	143.32	$KHC_8H_4O_4$ (phthalate)	204.224
Ag_2CrO_4	331.73	$KH(IO_3)_2$	389.912
$Al_2(SO_4)_3 \bullet 14(H_2O)$	594.35	K_2HPO_4	174.176
$BaSO_4$	233.39	KH_2PO_4	136.086
$CaCO_3$	100.089	$KHSO_4$	136.16
CaC_2O_4	128.100	KI	166.003
CaF_2	78.077	KIO_3	214.001
CaO	56.079	KIO_4	230.000
$Ca(OCl)_2$	142.985	$KMnO_4$	158.034
$Ca(OH)_2$	74.095	KNO_3	101.101
CO_2	44.010	$Mg(OH)_2$	58.320
Cl_2	70.906	$MgSO_4$	120.36
ClO_2	67.452	MnO_2	86.937
Cr_2O_3	151.990	Mn_2O_3	157.874
CuO	79.545	Mn_3O_4	228.812
$CuSO_4$	159.60	$NaBr$	102.894
$NHCl_2$	85.921	$NaCl$	58.443
$FeCl_3$	162.206	$NaCN$	49.007
FeO	71.846	Na_2CO_3	105.989
Fe_2O_3	159.692	$Na_2Al_2O_4$	163.940
Fe_3O_4	231.539	$NaHCO_3$	84.007
$Fe_2(SO_4)_3$	399.87	NaF	41.988
$Fe_2(SO_4)_3 \bullet 7(H_2O)$	525.97	$NaOH$	39.997
HBr	80.912	$Na_2O \bullet (SiO_2)_x$	Variable
$HC_2H_3O_2$ (acetic acid)	60.052	Na_2SiF_6	188.056
HCO_3^-	61.017	NCl_3	120.366
HCl	36.461	NH_3	17.030
$HClO_4$	100.458	NH_2Cl	51.476
HNO_3	63.013	NH_4Cl	53.491
H_2O	18.015	NH_4NO_3	80.043
H_2O_2	34.015	$(NH_4)_2SO_4$	132.13
H_3PO_4	97.995	NO_3^-	62.005
H_2S	34.08	O_3 (ozone)	47.998
H_2SO_3	82.07	OH^-	17.007

*Based on US Government Printing Office of atomic weights from Table C-1 above.

Compound	Weight in Grams	Compound	Weight in Grams
H_2SO_4	98.07	OCl^-	51.452
H_2SiF_6	144.092	$PbCrO_4$	323.2
HgO	216.59	$Pb(NO_3)_2$	331.2
Hg_2Cl_2	472.09	PbO	223.2
$HgCl_2$	271.50	PbO_2	239.2
KBr	119.002	$PbSO_4$	303.3
$KBrO_3$	167.000	P_2O_5	141.944
KCl	74.551	Sb_2S_3	339.68
$KClO_3$	122.550	SiO_2	60.084
KCN	65.116	$SnCl_2$	189.596
K_2CrO_4	194.190	SnO_2	150.689
$K_2Cr_2O_7$	294.184	SO_2	64.06
$K_3Fe(CN)_6$	329.248	SO_3	80.06
$K_4Fe(CN)_6$	368.346	$Zn_3(PO_4)_2$	304.703

APPENDIX D

DEPTH/DIAMETER (D/D) TABLE

LESS THAN FULL PIPELINE FLOW

\ TABLE D-1 Depth/Diameter (d/D) Table							
d/D	**Factor**	**d/D**	**Factor**	**d/D**	**Factor**	**d/D**	**Factor**
0.01	0.001	0.26	0.162	0.51	0.403	0.76	0.641
0.02	0.004	0.27	0.171	0.52	0.413	0.77	0.649
0.03	0.007	0.28	0.180	0.53	0.423	0.78	0.657
0.04	0.011	0.29	0.189	0.54	0.433	0.79	0.666
0.05	0.015	0.30	0.198	0.55	0.443	0.80	0.674
0.06	0.019	0.31	0.207	0.56	0.453	0.81	0.682
0.07	0.024	0.32	0.217	0.57	0.463	0.82	0.689
0.08	0.029	0.33	0.226	0.58	0.472	0.83	0.697
0.09	0.035	0.34	0.236	0.59	0.482	0.84	0.704
0.10	0.041	0.35	0.245	0.60	0.492	0.85	0.712
0.11	0.047	0.36	0.255	0.61	0.502	0.86	0.719
0.12	0.053	0.37	0.264	0.62	0.512	0.87	0.725
0.13	0.060	0.38	0.274	0.63	0.521	0.88	0.732
0.14	0.067	0.39	0.284	0.64	0.531	0.89	0.738
0.15	0.074	0.40	0.293	0.65	0.540	0.90	0.745
0.16	0.081	0.41	0.303	0.66	0.550	0.91	0.750
0.17	0.089	0.42	0.313	0.67	0.559	0.92	0.756
0.18	0.096	0.43	0.323	0.68	0.569	0.93	0.761
0.19	0.104	0.44	0.333	0.69	0.578	0.94	0.766
0.20	0.112	0.45	0.343	0.70	0.587	0.95	0.771

d/D	Factor	d/D	Factor	d/D	Factor	d/D	Factor
0.21	0.120	0.46	0.353	0.71	0.596	0.96	0.775
0.22	0.128	0.47	0.363	0.72	0.605	0.97	0.779
0.23	0.137	0.48	0.373	0.73	0.614	0.98	0.782
0.24	0.145	0.49	0.383	0.74	0.623	0.99	0.784
0.25	0.154	0.50	0.393	0.75	0.632	1.00	0.785

APPENDIX E

The following 12 diagrams show some of the most common types of wastewater treatment plants with the last diagram showing sludge processing. Wastewater plants shown in figures 9 and 10 are newer technologies using membranes, nanofiltration, and electrodialysis.

Please note that not all the processes in any diagram are necessarily used. Also, some steps are not depicted. There are of course many more wastewater treatment plant types and process arrangements that are not shown. The number of different wastewater treatment plant arrangements is beyond the scope of this book. For further study of other wastewater treatment plants, please see the references.

WASTEWATER TREATMENT AND SLUDGE PROCESSING FLOW CHARTS

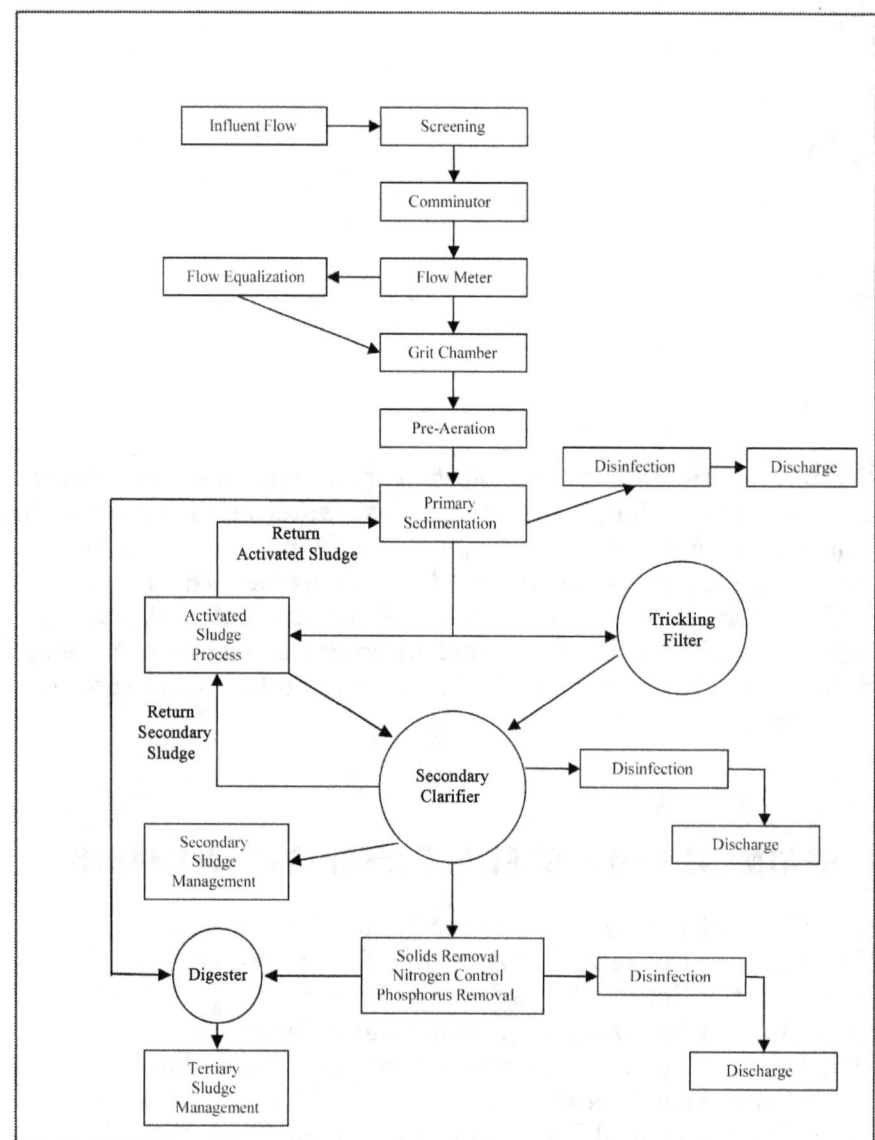

Figure 1 Flow chart of typical wastewater treatment processes

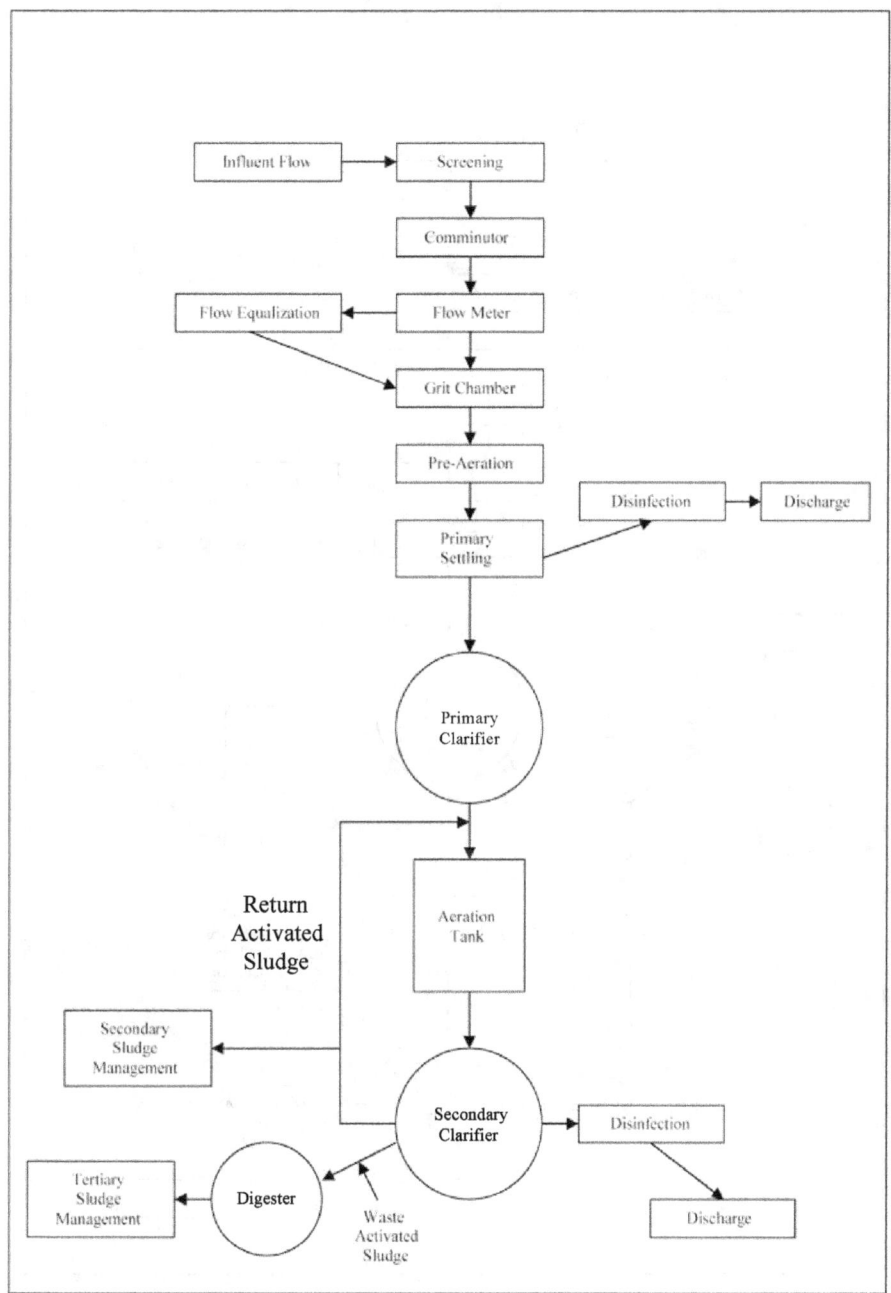

Figure 2 Flow chart of conventional activated sludge process

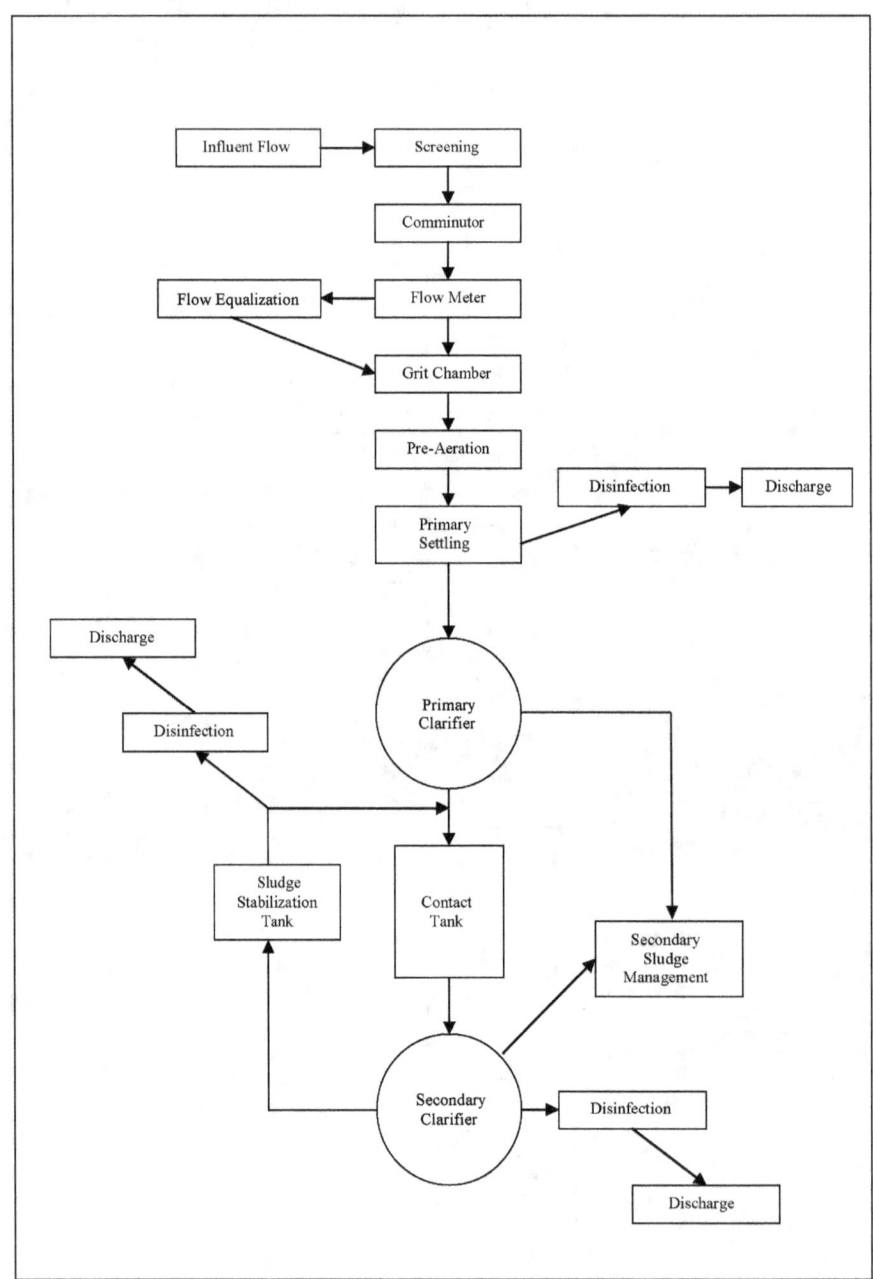

Figure 3 Flow chart of contact stabilization process

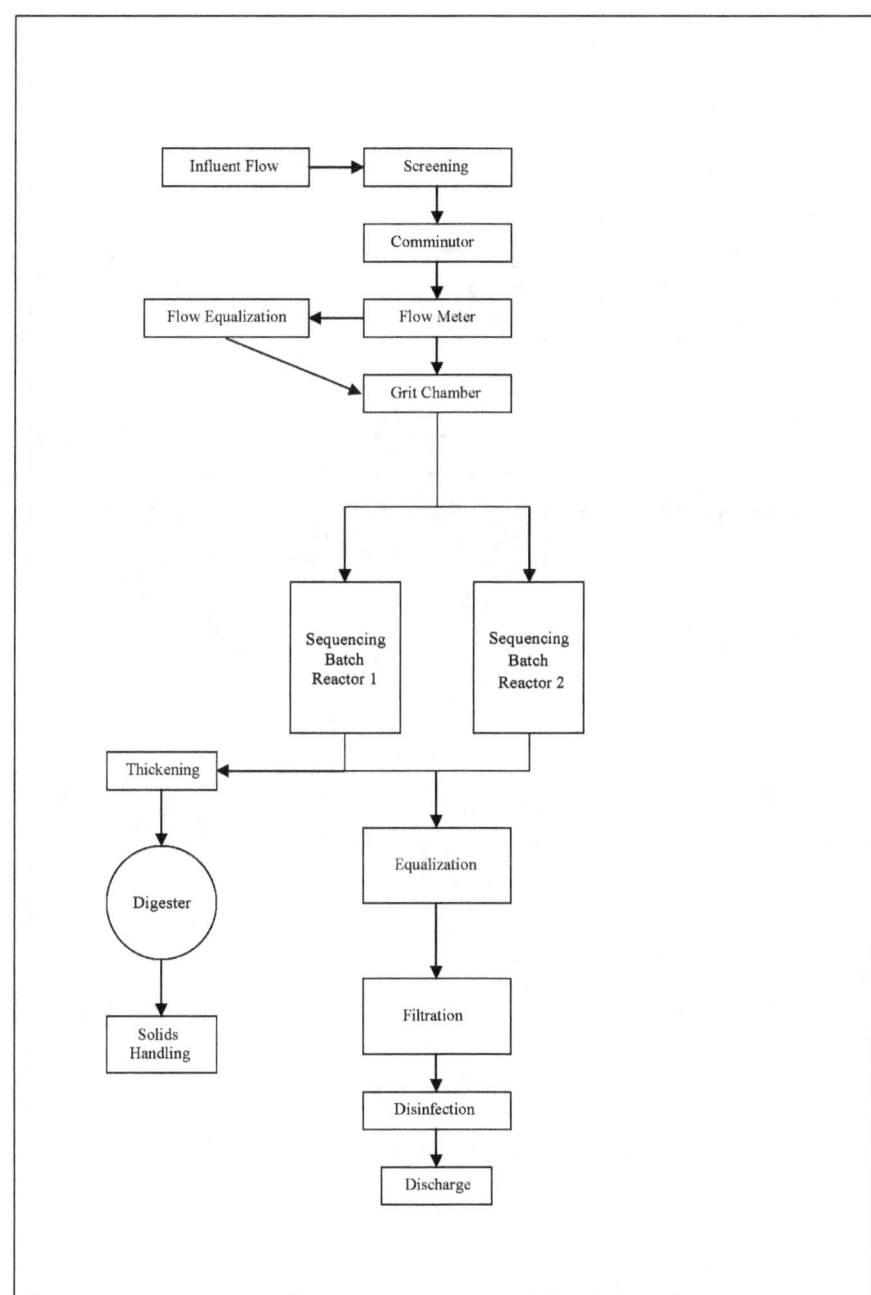

Figure 4 Flow chart of activated sequencing batch reactor

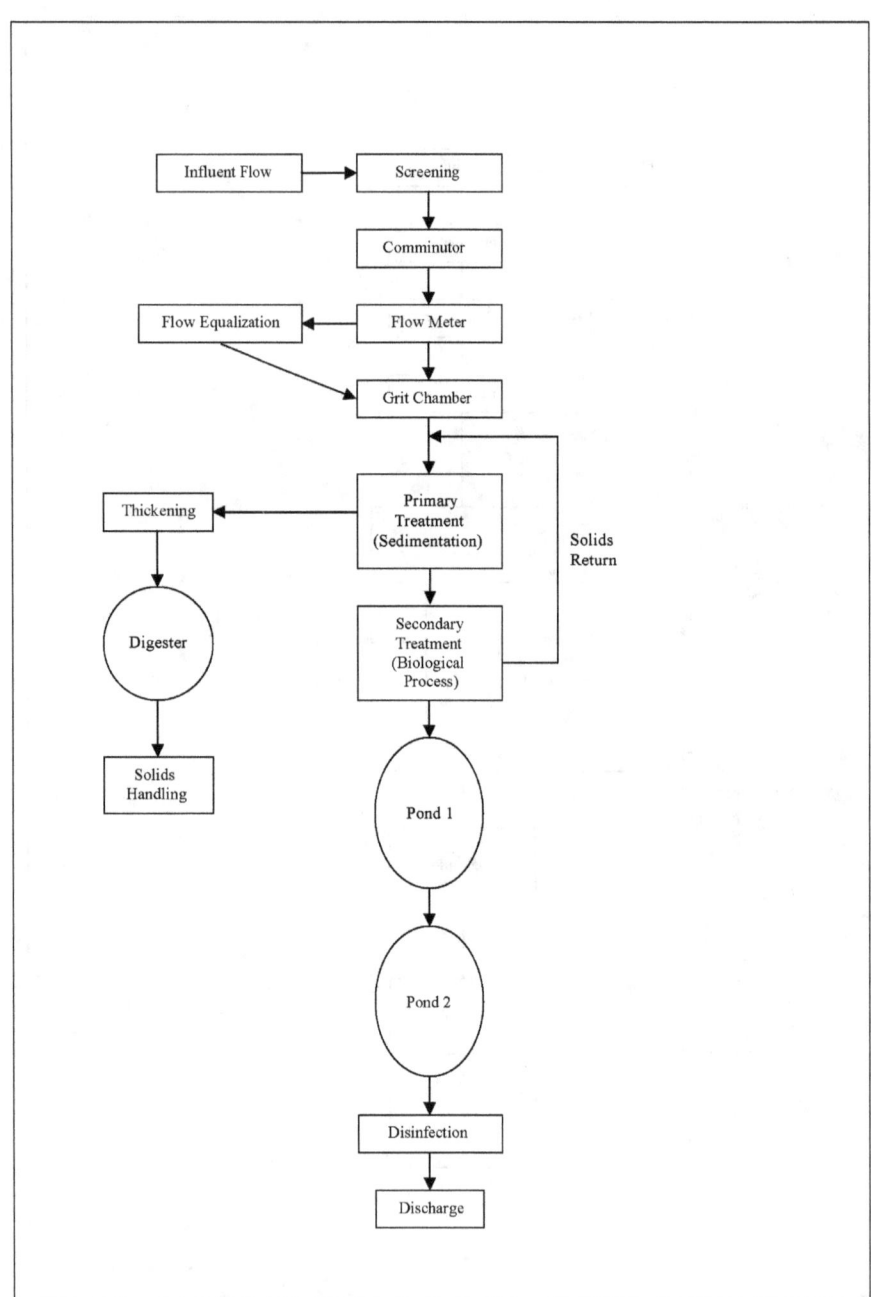

Figure 5 Flow chart of typical wastewater treatment using ponds after secondary treatment

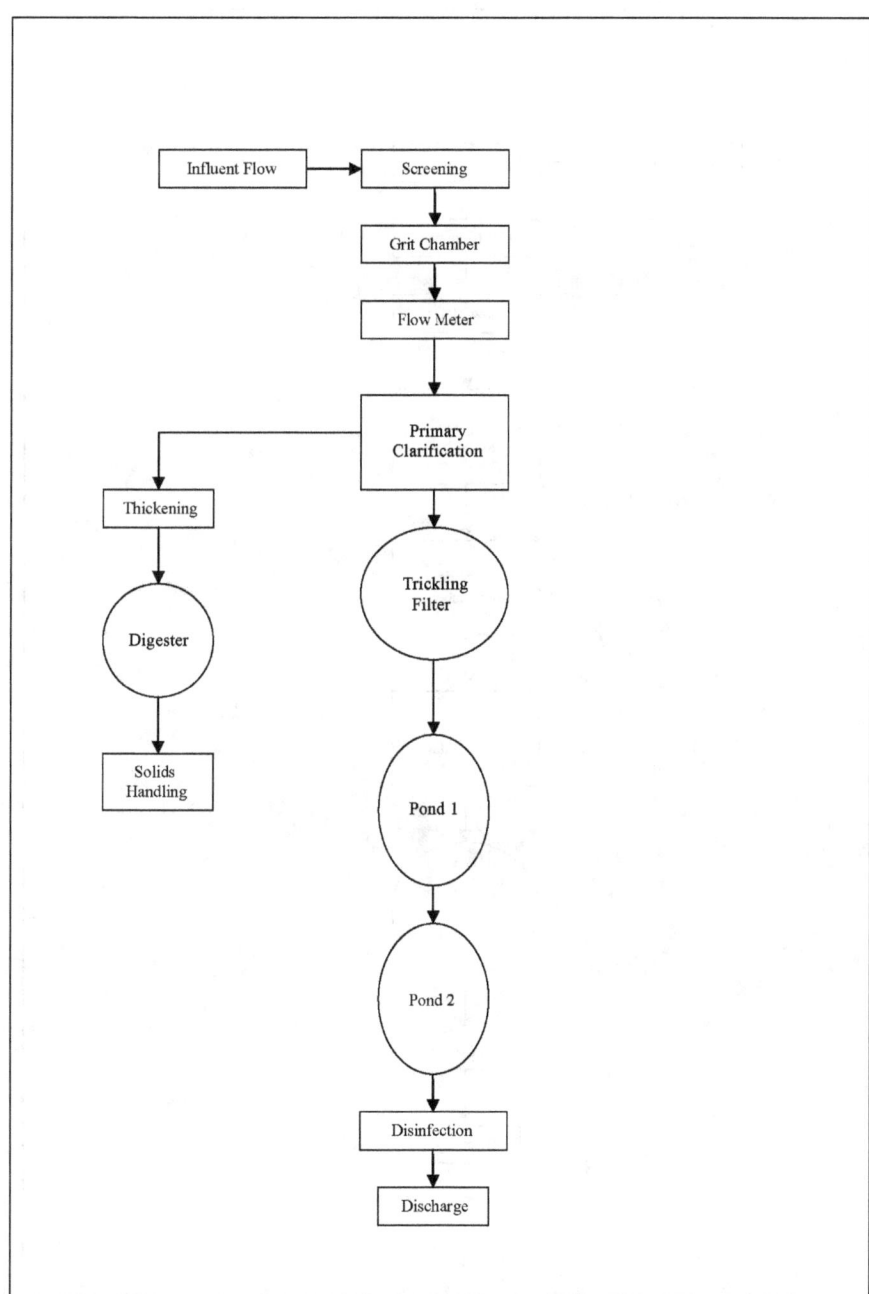

Figure 6 Flow chart of typical wastewater treatment using polishing ponds in series treatment

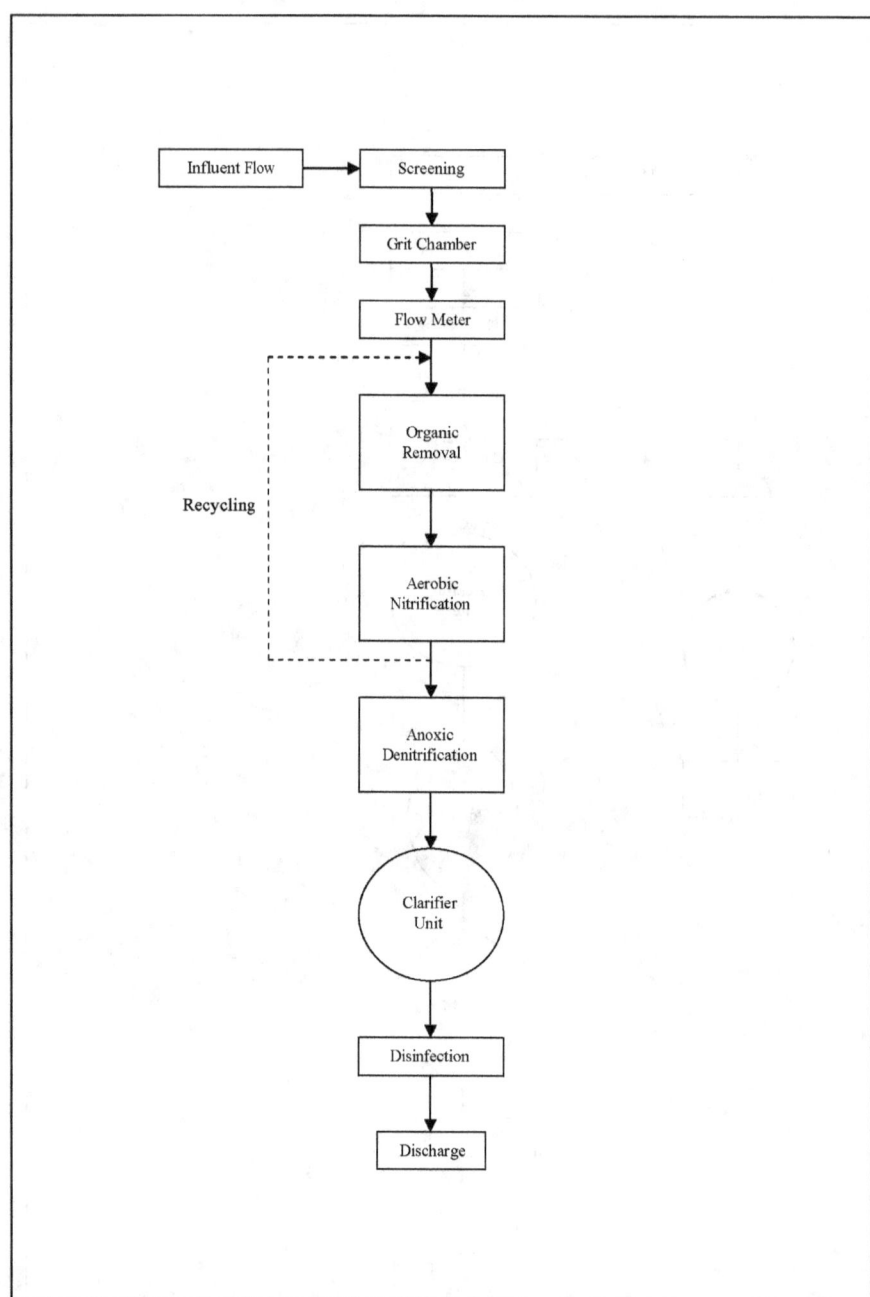

Figure 7 Flow chart of wastewater treatment using rotating biological contactor process

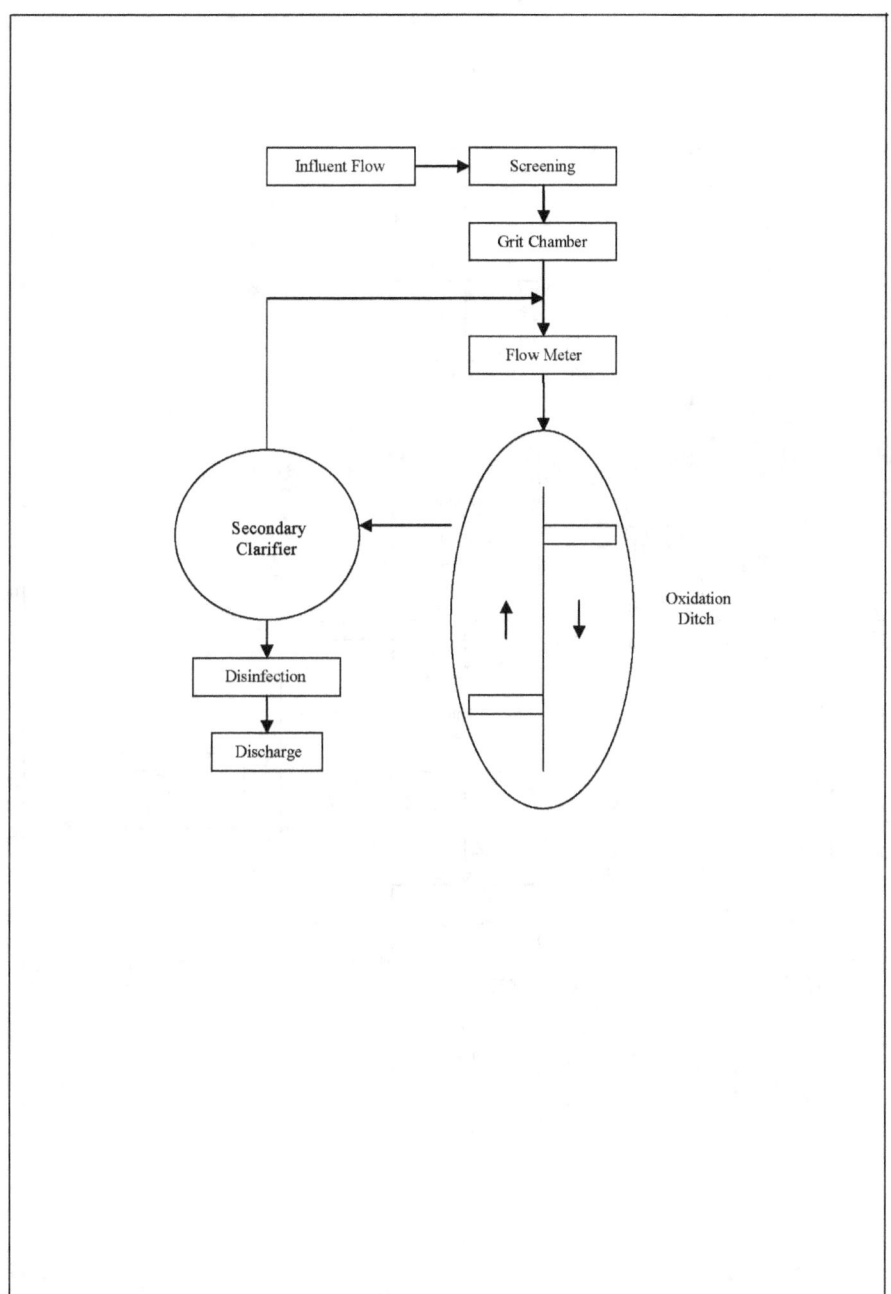

Figure 8 Flow chart of typical wastewater treatment using an oxidation ditch

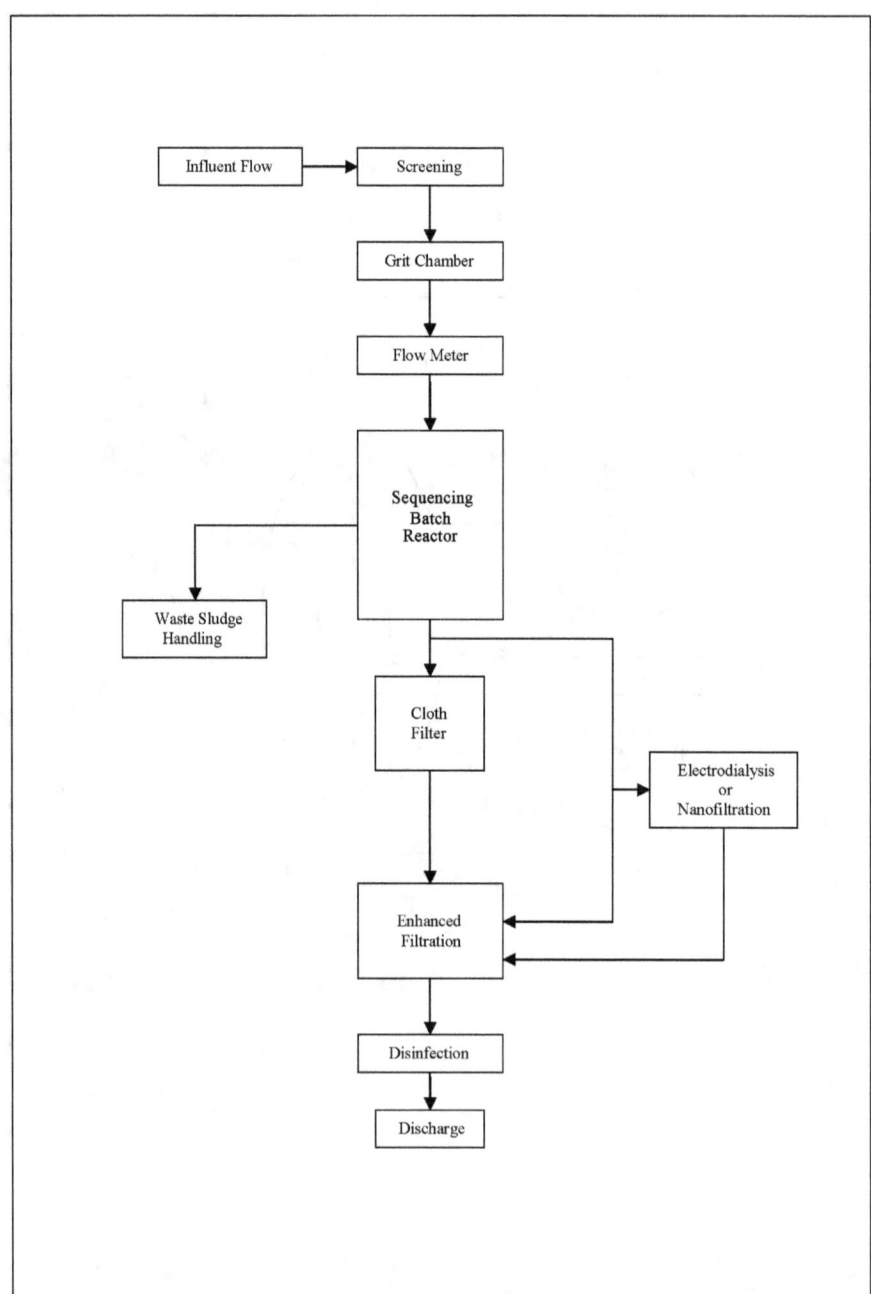

Figure 9 Flow chart of typical wastewater treatment using electrodialysis or nanofiltration

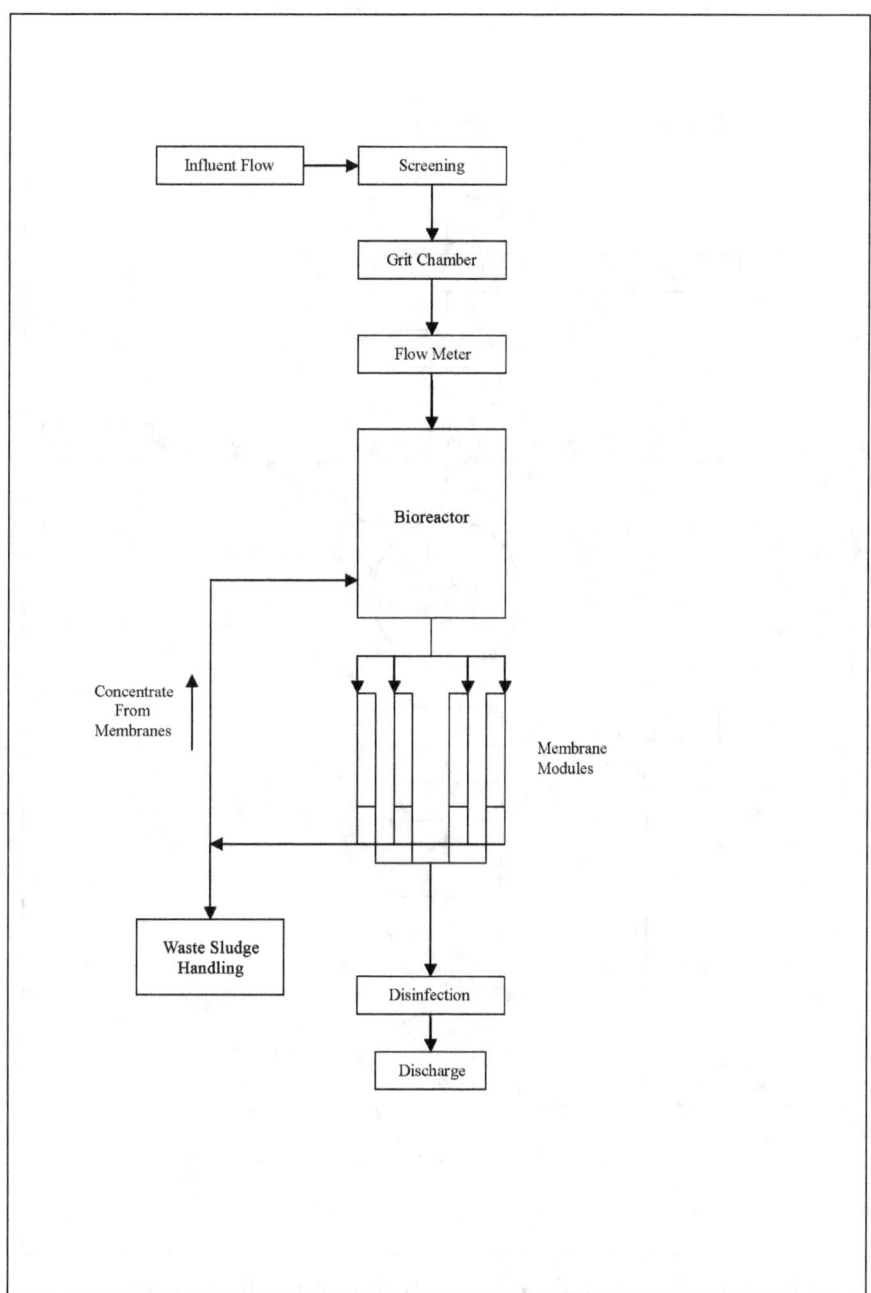

Figure 10 Flow chart of typical wastewater treatment using a membrane bioreactor

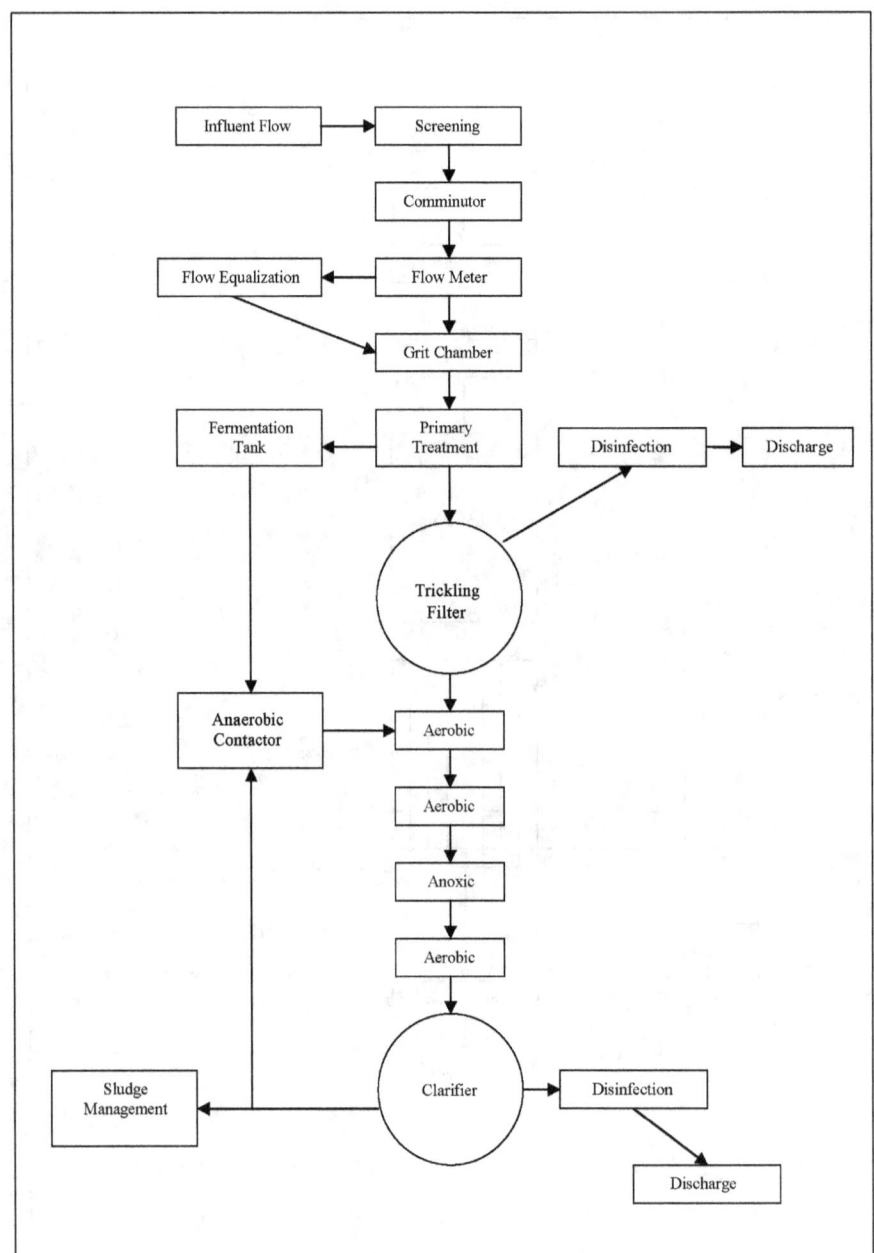

Figure 11 Flow chart of wastewater treatment nitrification process

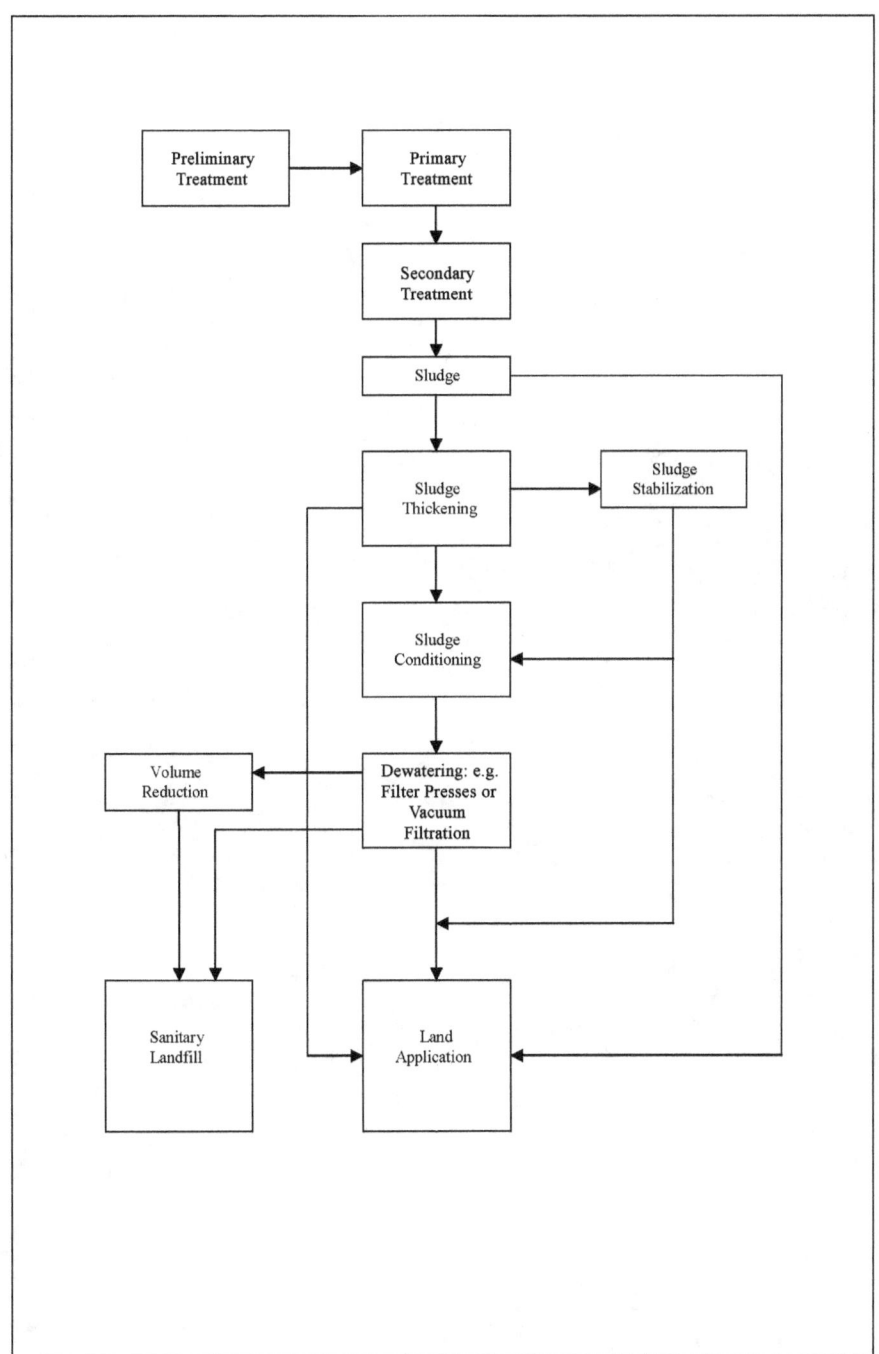

Figure 12 Flow chart of wastewater treatment plant processing sludge

APPENDIX F

acre-ft	acre-feet
amps	amperes
AT	Aeration Tank
avail.	available
avg.	average
BC	Blending Compost
bhp	brake horsepower
BOD	Biochemical Oxygen Demand
°C	Degrees Centigrade
COD	Chemical Oxygen Demand
cm	centimeter(s)
CT	Clarifier Tank
D	Diameter or Depth (Note context)
d	day
DAF	Dissolved Air Flotation
DB	Dewatered Biosolids
DO	Dissolved Oxygen
effic.	efficiency
°F	degrees Fahrenheit
F/M	Food-to-Microorganism ratio
ft	foot or feet
ft/s	feet per second
ft^2	square feet
ft^3	cubic feet
f^3/min	cubic feet per minute

ft³/s	cubic feet per second
g	gram(s) or gravity (Note context)
gal	gallon(s)
gpcpd	gallons per capita per day
gpd	gallons per day
gph	gallons per hour
gpm	gallons per minute
gr-eq	gram equivalent weight
hp	horsepower
hr	hour(s)
HTH	High Test Hypochlorite
in.	inch(es)
kg	kilogram
kW	Kilowatt(s)
lb	pounds
L	Liter or Length (Note context)
m	meter
m³	cubic meters
M	Mole(s)
mA	milliamp(s)
MCRT	Mean Cell Residence Time
MCL	Maximum Contaminant Level
ME	Motor Efficiency
meq	milliequivalent
mg	milligrams
mgd	million gallons per day
mg/L	milligrams per liter
mhp	motor horsepower
mil	million
mil gal	million gallons
min	minute
mL	milliliter
MLSS	Mixed Liquor Suspended Solids
MLVSS	Mixed Liquor Volatile Suspended Solids
MR	Mix Ratio
N	Normality
ntu	nephelometric turbidity units
oz	ounce(s)

pH	Hydrogen ion concentration
PE	Pump Efficiency
%	percent
ppm	parts per million
psi	pressure per square inch absolute
psig	pressure per square inch gauge
Q	Flow
r	radius
RAS	Return Activated Sludge
RBC	Rotating Biological Contactor
s	second(s)
sec	second(s)
sed	sedimentation
soln.	Solution
sp gr	specific gravity
SRT	Solids Retention Time
SS	Suspended Solids
SSC	Sludge Solids Concentration
temp.	temperature
TDH	Total Dynamic Head
TDS	Total Dissolved Solids
TS	Total Solids
TSS	Total Suspended Solids
TH	Total Head
V	Volume or Velocity or Volt(s) Note context
VR	Volatilization Rate
VS	Volatile Solids
VSC	Volatile Solids Concentration
VSR	Volatile Solids Reduction
W	Width
w	watt
WAS	Waste Activated Sludge
whp	water horsepower
wt	weight
WVS	Waste Volatile Solids
yr	year
yd^3	cubic yards

BIBLIOGRAPHY

American Water Works Association. *AWWA Wastewater Operator Field Guide*. Denver, Colo.: American Water Works Association, 2006.

American Water Works Association. *Basic Science Concepts and Applications*. 3rd ed. Denver, Colo.: American Water Works Association, 2003.

Boikess, Robert S. *How to Solve General Chemistry Problems*. 8th ed. Englewood Cliffs, N.J.: Prentice Hall Inc., 2008.

Forster, Christopher. *Wastewater Treatment and Technology*. London: Thomas Telford Limited, 2003.

Frey, Paul R. *Chemistry Problems and How to Solve Them*. 8th ed. Barnes and Noble College Outline Series. New York: Barnes and Noble Inc., 1985.

Giorgi, John. *Math for Water Treatment Operators: Practice Problems to Prepare for Water Treatment Operator Certification Exams*. Denver, Colo.: American Water Works Association, 2007.

Giorgi, John. *Math for Distribution System Operators: Practice Problems to Prepare for Distribution System Operator Certification Exams*. Denver, Colo.: American Water Works Association, 2007.

Idaho Department of Environmental Quality. *Wastewater Land Application Operators Study and Reference Manual*. Boise, Idaho: Idaho Department of Environmental Quality, 2005.

Lin, Shun Dar. *Water and Wastewater Calculations Manual*. 2nd ed. New York: McGraw-Hill, 2007.

Operation of Wastewater Treatment Plants, Volume 1: A Field Study Training Program. 6th ed. Sacramento, Calif.: California State University, Sacramento School of Engineering, 2004.

Operation of Wastewater Treatment Plants, Volume 2: A Field Study Training Program. 6th ed. Sacramento, Calif.: California State University, Sacramento School of Engineering, 2003.

Price, Joanne Kirkpatrick. *Applied Math for Wastewater Plant Operators*. Boca Raton, Fla.: CRC Press, 1998.

Price, Joanne Kirkpatrick. *Basic Math Concepts for Water and Wastewater Plant Operators*. Lancaster, Pa: Technomic Publishing Co., 1991.

Skoog, Douglas A., Donald M. West, F. James Holler, and Stanley R. Crouch. *Fundamentals of Analytical Chemistry*. 8th ed. Pacific Grove, Calif.: Brooks Cole, 2003.

Spellman, Frank R. *Mathematics Manual for Water and Wastewater Plant Operators*. New York: CRC Press, 2004.

ADDITIONAL RESOURCES

BASIC CHEMISTRY FOR WATER AND WASTEWATER OPERATORS

A basic chemistry primer tailored for operators of drinking water or wastewater systems.

Edition: 2002, Softbound, 178 pp.
ISBN 1-58321-148-9; Catalog Number 20494

BASIC MICROBIOLOGY FOR DRINKING WATER PERSONNEL

This book provides clear, short descriptions of the waterborne microorganisms—viruses, bacteria, protozoa, and algae—that either pose a human health threat or contribute to distribution corrosion.

Edition: 2001, Softbound, 85 pp.
ISBN 1-58321-121-7; Catalog Number 20463

ESSENTIAL WATER AND WASTEWATER CALCULATIONS FOR ENGINEERS AND OPERATORS

This book contains hundreds of water and wastewater engineering calculation procedures ranging from simple to complex. Each calculation provides easy-to-follow steps for solving and an example that demonstrates important concepts.

Edition: 2007, Hardback, 372 pp.
ISBN 0-97456-898-8; Catalog Number 20656

WASTEWATER MICROBIOLOGY: A HANDBOOK FOR OPERATORS

Wastewater treatment is a microbiological process. Microorganisms such as bacteria and protozoa do the actual breakdown and removal of nutrients and organic material in the wastewater. A wastewater treatment plant operator's job is to control this biological process. That is why wastewater operators need to understand basic microbiology, as well as the types of microorganisms that are used in the treatment of sewage, and how the microbes do their job in the wastewater treatment process.

Chapters cover wastewater treatment system overview, general microscopy, bacteria, protozoa, metazoans, filamentous bacteria, and microbiology and process control. Includes glossary of terms and CD-ROM with 85 color photographs of microorganisms.

Edition: 2005, Softbound, 182 pp.
ISBN 1-58321-343-0; Catalog No. 20563

For pricing and ordering information, please visit the online bookstore at www.awwa.org/bookstore or call AWWA Customer Service at 1.800.926.7337.

www.ingramcontent.com/pod-product-compliance
Lightning Source LLC
Chambersburg PA
CBHW081104170526
45165CB00008B/2322